Nine Steps from Ignition to Extinguishment

Second Edition

FIREPRO® *Institute Ltd.*   Putney, Vermont USA

Cover Photographs:

Top Left and Lower Right - A major modern city.

Lower Left - A 31-story building 25 minutes after ignition, before manual firefighting
starts ... the largest flame yet designed (unintentionally) in a building.

Upper Right - Conditions on the fireground are unpredictable and can be frustrating.

FIREPRO Institute Ltd., Postal Box 600, Putney, Vermont 05346, USA

Early Published Works 1975
Published 1976. Second Edition 1994
Printed in the United States of America
98 97 96 95 94   10 9 8 7 6 5 4 3 2 1

Wilson, Rexford.
        Nine steps from ignition to extinguishment.
        Rexford Wilson. 1. Time in Fire Control. 2. Fire Extinguishment.
3. Fire Termination. 4. Fire Service Time. 5. M-Curve Quantification. I. Title

Library of Congress Catalog Card Number: 94-90840

ISBN 1974496791

Design and imagesetting by Joy Wallens, PenTangle Studio, Brattleboro, Vermont

Printed on recycled paper by BookMasters, Inc., Ashland, Ohio

Nine Steps from Ignition to Extinguishment

Second Edition

Rexford Wilson
FSFPE

Illustrations by
Donald W. Demers

FIREPRO® *Institute Ltd.* Putney, Vermont USA

Nine Steps from Ignition to Extinguishment

by Rexford Wilson, FSFPE

Abstract

In every fire incident there is a measurable continuum of time from the moment of *ignition* through the moment of *extinguishment*.

This publication defines a series of **9 Steps** along the incident time continuum for fires in structures. Each Step occurs between two **Points** in time. This sequence of carefully selected, observable points is: Ignition, Recognition, Detection, Alarm, Alert, Get-Out, Arrival, Agent Application, Flameout, and Extinguishment.

In addition to definitions for these **Points** and the **9 Steps** between them, discussions follow that present examples from the field to illustrate factors influencing the length of each Step. Definitions for three commonly noted but opinion-based *Substeps* of the **Combat Step** are included for those who wish to use them.

The overall time continuum of a fire incident is also segmented into two **Key Times**, *Reflex Time* and *Subdue Time*, which are separated by the "magic moment" of Agent Application. The three fire **Management Periods** (*Pre-Response, Response,* and *Incident*) are "controlled" by three different authorities: the property owner, the Fire Service's chief officer, and the Fire Service's incident commander.

The section on timekeeping for analyzing a fire incident in terms of the **9 Steps** shows how information gathered from many incidents can be routinely applied to shorten the overall time required for future fire incident responses.

The terminology includes words used in various areas of the United States. Readers using different languages or from other areas of the USA are encouraged to substitute the terms they use by writing them in the margin.

While this document is based on fires in enclosures, such as buildings or other structures, readers dealing with outdoor fires or with other types of emergencies should feel free to adapt the text to their needs.

Dedication

To the men and women of the Fire Service -- worldwide,
for their continued effort to preserve life and property.

Acknowledgments

Many, many persons have helped shape this document with ideas, discussion, and encouragement. Among those are a great number with experience in Fire Services around the globe, including:

Byron Chaney, *former chief of department in Mountainview, California, USA*
C. Walter Stickney, *former state fire marshal of Oregon, USA*
Russell Sanders, *chief of department in Louisville, Kentucky, USA*
Lyle Goodrich, *former training officer for the state of Washington, USA*
Charlie Rule, *chief of department in Manteca, California, USA*
Martin Grimes, *former fire department commissioner in Bermuda*
Ed Herron, *former state fire marshal of Iowa, USA*
John Bunting, *chief of department in New Boston, New Hampshire, USA*
Ernie Emerson, *former state fire marshal of Texas, USA*
Gert Magnus, *former chief of department, Mannheim, Germany*
Terry Hayes, *former fire marshal of Shreveport, Louisiana, USA*
Curt Volkamer, *former chief of department in Chicago, Illinois, USA*
Tom Brace, *state fire marshal of Minnesota, USA*
Lewis Burton, *former chief of department in Haverhill, Massachusetts, USA*
Bill Miller, *former chief of department in Los Angeles, California, USA*
Tom McNerney, *former training officer in Seattle, Washington, USA*
Milt Bullock, *former chief of department in Dade County, Florida, USA*
Cliff Harvey, *assistant chief of department in Boulder, Colorado, USA*
Henry Thomas, *former chief of department in Hartford, Connecticut, USA*
Ron Coleman, *state fire marshal of California, USA*
Delvin R. Bunton *of the USDA Forest Service in Portland, Oregon, USA*
George Paul, *former chief of department in Boston, Massachusetts, USA*
Steve Black, *chief of department in Peterborough, New Hampshire, USA*

Others outside the Fire Service also made valuable contributions, especially to establishing the need for this document. These include:

Dr. Robert W. Fitzgerald, *Worcester Polytechnic Institute, Worcester, Mass., USA*
Dr. David Rasbash, *University of Edinburgh, Scotland*
Mr. Christopher L. Wilson, *Wellesley, Massachusetts, USA*
Dr. John Bryan, *University of Maryland, College Park, Maryland, USA*
Dr. R. Brady Williamson, *University of California, Berkeley, California, USA*

To all those who have helped with the evolution of the 9 Steps concept, I extend my thanks.

Special thanks go to my wife Margaret for her patience and moral support during the development of this and other manuscripts.

I am also indebted to my associate Joy Wallens, who offered wise suggestions regarding the text's content and presentation, and in whose hands the *Nine Steps* material has been crafted into its current form.

Donald W. Demers, who illustrated the first edition, is now a widely noted illustrator and cover artist. His artwork is again used here as a tribute to him.

Contents

Foreword

Manual extinguishment of fires by the Fire Service traditionally has been a significant part of building and community firesafety. In many cases, the Fire Service is the only active defense an individual or building may have for protection from the ravages of unwanted, hostile fires. Over the years, the Fire Service has proven its resourcefulness and ingenuity in improving the effectiveness of the activities it delivers.

In this short primer that describes the process, Rexford Wilson has provided a blueprint that can be used for a variety of purposes. As one important function, this document provides a tool that can be used to communicate Fire Service needs and the importance of time in the delivery of services to the public and to owners and occupants of buildings. Such communication can also provide a basis for working with the building designer in order to produce buildings that are designed for fire suppression to a greater degree than now occurs. The effectiveness of this communication is enhanced by identification of a vocabulary that describes the events and time intervals in a clear, concise manner. This feature alone makes the document valuable.

As another major contribution, the Fire Service is provided with a documentation of the entire process from ignition through notification to agent application and, finally, extinguishment and overhaul. This understanding of the entire process and the incorporation of illustrative examples may provide an opportunity for the Fire Service itself to reconsider its standard procedures in order to cut those precious seconds to reduce damage and save lives.

Finally, this short document provides a clear understanding of the type of statistical data that is needed to provide measurements of performance. The measurements of performance relate not only to the delivery of services by a modern fire department, but also to the way in which community decision making affects those services. In fire design, it is often noted that the building design for fire suppression should measure the building's ability to allow the Fire Service to accomplish its task. *Nine Steps from Ignition to Extinguishment* provides the framework from which we can identify the important data for comparing buildings and building features in a manner clearly perceived by layman and professional alike.

Robert W. Fitzgerald
Worcester Polytechnic Institute

October 18, 1994

FIREPRO® Institute Ltd.

Preface

Fire power can grow rapidly; it doubles and doubles again. Fire grows on a mathematical (exponential) scale, not on a human (linear) scale. Fast fires kill more people. Future performance-based Codes will require that we measure the effect of fire growth and our human reaction to it.

The basic firesafety objective of a Fire Service in building fires is to remove people from the fire's path and put the fire out.

The other tasks to help its population are secondary to this basic objective. Non-fire rescue, emergency medical, or hazardous materials incidents, while important, are not as time-sensitive since they tend to grow in a linear fashion. The pre-fire tasks of population and building fire readiness are generally not time-sensitive operations.

SUGGESTIONS FOR
REVIEWING THIS
DOCUMENT:

1. *Flip pages to get a general sense of the layout.*

2. *Change the terminology and units to your own.*

3. *Study the parts you wish.*

Every Fire Service jurisdiction provides six functions: communications, operations, prevention, training, equipment, and administration. Each of these six functions is geared toward preparing the population, personnel or equipment for the basic firesafety objective: remove the people, put out the fire. This is the job of Manual Fire Suppression.

In order to measure the effectiveness of Manual Fire Suppression, accurate records of certain intervals of time use are needed. To obtain this timing, accurate definitions of the time intervals or "**Steps**" are required. This document is produced for that purpose.

Words are important if we are to communicate effectively together worldwide. The word "Realms" is used in Fire Growth. "Episodes" is used in Human Action. "Steps" is used in Fire Suppression.

The ideas are not just the author's; contributions have come from many sources. Each of the "**Steps**" has been improved. Terminology has been adjusted to match the Measurement of Building Firesafety method — the L-Curve method. Standard SI units for firesafety are followed with US units in parentheses, e.g., "1 kilometre (0.6 miles)." Fill in the terms and units you use to think with — the margins are for that purpose.

Shaving 10 seconds off any point in the linear series of "**Steps**" before the **Agent Application Point** will cause agent to be applied 10 seconds earlier. Each action taken to reduce the **Reflex Time** will reduce the size of the fire attacked, save stress on those affected, reduce the risk to those trying to handle the situation, and lower the loss of life or property.

One student of the previous edition has called the work "timeless." Enjoy these thoughts. When you improve them, please let the author know through the FIREPRO Institute Ltd. for use in the next edition.

Rexford Wilson

Introduction

It's typical:

The officer of the first-arriving piece of fire apparatus steps out. A witness runs up and shouts, "It took you 15 minutes to get here!" The fire officer knows that he left the station and got to the scene within 3 minutes. How can this difference be explained?

Perhaps both times are correct, and each person is actually measuring a different period of time. The witness may be timing from the moment he first became aware of the fire to the moment he first became aware of the engine's arrival. The fire officer may be timing from when he left the station to the time he could first see the incident building.

Managing Time

In order to manage the time required for Fire Service attack and extinguishment of fires, a common set of benchmarks is needed. These benchmarks can be thought of as a series; that is, one Step is completed before the next Step is begun. This series can be reduced to **9 Steps** in each fire between the point of initial ignition and the point of final extinguishment.

Steps

There is a period of time, or "step," that takes place between each of two defined "points." The **"Travel"** Step (time) may be very short in one incident, but the **"Setup"** Step in that same incident may be very lengthy. The next incident, however, may involve a long **"Travel"** Step and a very easy access and **"Setup"** Step.

In order to make sense of the normal way manual fire suppression occurs, it is necessary to define certain steps that always occur. An effort has been made to further define specific "points" that always occur at the beginning and ending of each step. The steps, then, can be uniformly recorded for meaningful analysis and action.

Actions taken by building occupants or passersby are not included here.

Points

Each fire spans many points or "seconds" in time. A point can be defined as a one-second moment in time. The points used here, however, have been carefully selected to be physically observable and easily reportable. These selected points are "SMART" points — i.e., each is: Specific, Meaningful, Acquirable, Realistic and Timeable. This document uses SMART points that are based on physical facts rather than on opinions.

An example of a SMART "point" is *the front wheel of the first piece of Fire Service apparatus first stops in front of the incident building.* That is a one-second period in time; it is not an opinion of the company officer or of anyone else. It is a specific second in the history of humankind.

Another example of a SMART "point" might be *the first drop of water out of the Fire Service nozzle in a room with flame present.* This **Agent Application Point** is the most important point in the process. Again, this is not an opinion; it is a physical fact. It is one second in the history of humankind.

Key Times

There are two key times: **Reflex Time** and **Subdue Time**. The **Reflex Time** is the sum of the first 7 steps before the **Agent Application Point**. The **Subdue Time** is the sum of the remaining 2 steps after the **Agent Application Point**.

The **Agent Application Point** — the "magic moment" of Manual Fire Suppression — separates these two key times.

Time Capture

Points are easier to time. After definitions of the 9 Steps from ignition to extinguishment, you will find a discussion of the points involved that can be routinely collected. Included are examples of how time analysis has been used. Which data elements are to be collected, how they are collected, and the form they take, should be adapted to match local needs.

Time Analysis and Use

The collected times can be analyzed to guide management action.

Analyzing the time used in the 9 Steps in a Fire Service jurisdiction can develop both improved policy and procedures for:

- tactical use by the Fire Attack Team on an engine or a ladder;
- strategic planning and policy changes by the Fire Service Administrator; and
- personnel development by the Fire Service Trainer.

In addition, the information may help others, including:

- the Fire Engineer in making predictions before a fire;
- the Fire Pathologist in preparing an analysis after a fire;
- the Fire Researcher in the study of fire phenomena;
- City Planners in choosing future firesafety guidelines; and
- Equipment Designers in refining apparatus to fit the need.

Glossary

This document is focused on those incidents where fires are occurring *inside* buildings. The work can be adapted to fires occurring outside buildings and to other incidents. For the purposes of this document, the following terms are used:

Step — A step is an interval of time between two points.

Point — A specific moment in time.

Time — Time is the universal yardstick of fire control. For firesafety it is usually measured in seconds. The title "time" in this document has been reserved for **Reflex Time** and **Subdue Time**.

Ignition — The first open flame. Smoldering or non-flaming burning is important pre-ignition work.

Fuel — Anything that burns (e.g., wood, corrugated cardboard, plastic).

Fire Service — A unit of local government organized to extinguish unwanted fires inside buildings or outdoors. It may have other duties. The term is used regardless of training, equipping, and staffing of such an organization. In different areas of the world, the Fire Service is called different things — e.g., Fire Brigade, Fire Department.

Alarm Center — The Fire Service site where the initial alarm for a fire is processed. The Alarm Center may be in the local funeral parlor or in an organized Alarm Handling Facility. In any case, personnel in the Alarm Center decide how and when to alert which persons in the responding Fire Service.

Responding Station — The building or structure housing equipment required by the Fire Service to extinguish fires.

Incident — Any time the front wheels of a piece of Fire Service apparatus crosses the threshold of the Responding Station in response to an alarm, an incident occurs. The incident may be for a fire, medical, hazardous materials, or other call.

The "magic moment": the moment fire suppression actually begins; the moment of first agent application.

Agent Application Point — The first application of the first drop of extinguishing agent on the flames at the scene of a fire. For this document, the Agent Application Point will be restricted to the first Agent applied by Fire Service personnel.

Reflex Time — The number of seconds that occur between the Ignition Point and the first Agent Application Point.

Subdue Time — The number of seconds that occur between the Agent Application Point and the Extinguishment Point.

FIREPRO® Institute Ltd.

Chapter 1
The 9 Steps: An Overview

This chapter will give the reader:

- A picture of the **9 Steps** — that series of steps required for manual fire suppression by a Fire Service.

- A picture of the **Reflex Time** — the time <u>before</u> the **Agent Application Point**.

- A picture of the **Subdue Time** — the time <u>after</u> the **Agent Application Point**.

- A picture of the 3 separate Managers controlling the 9 Steps when each is in command.

- An insight into the importance of *a single second* in the overall consequences of a fire incident.

- A discussion of the term "response."

"When you cannot express it in numbers, your knowledge is of a meager and unsatisfactory kind." — Lord Kelvin

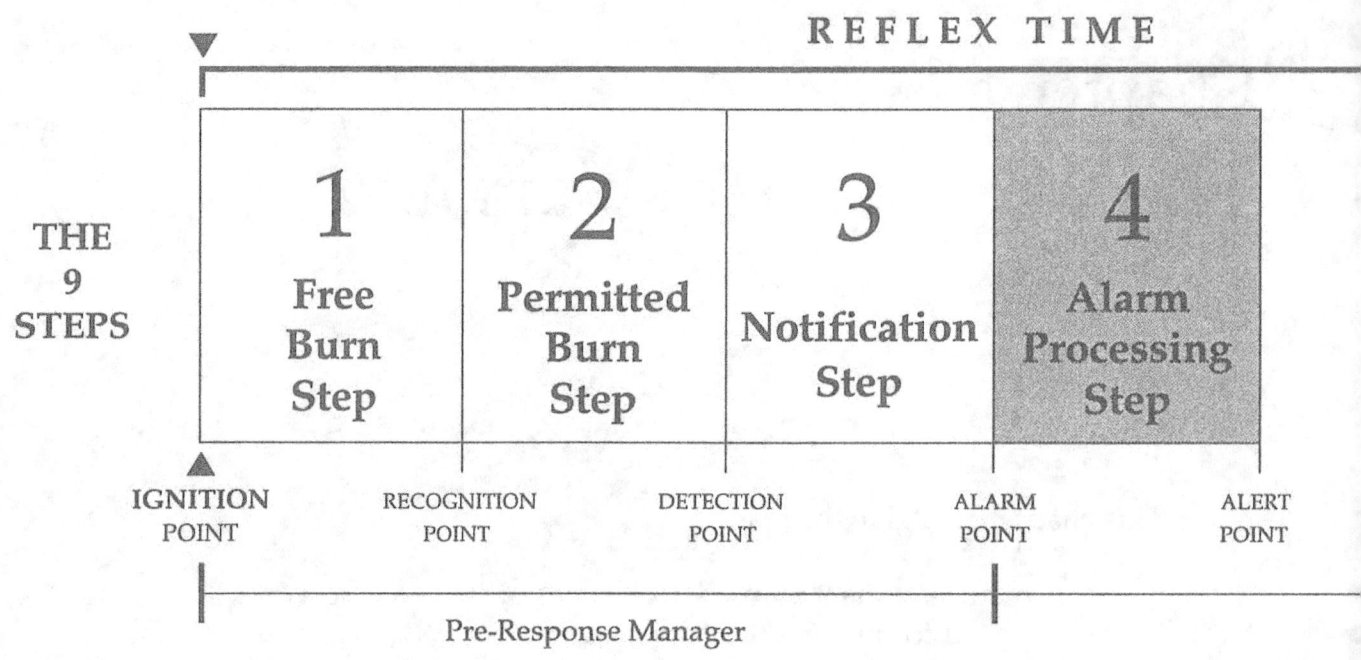

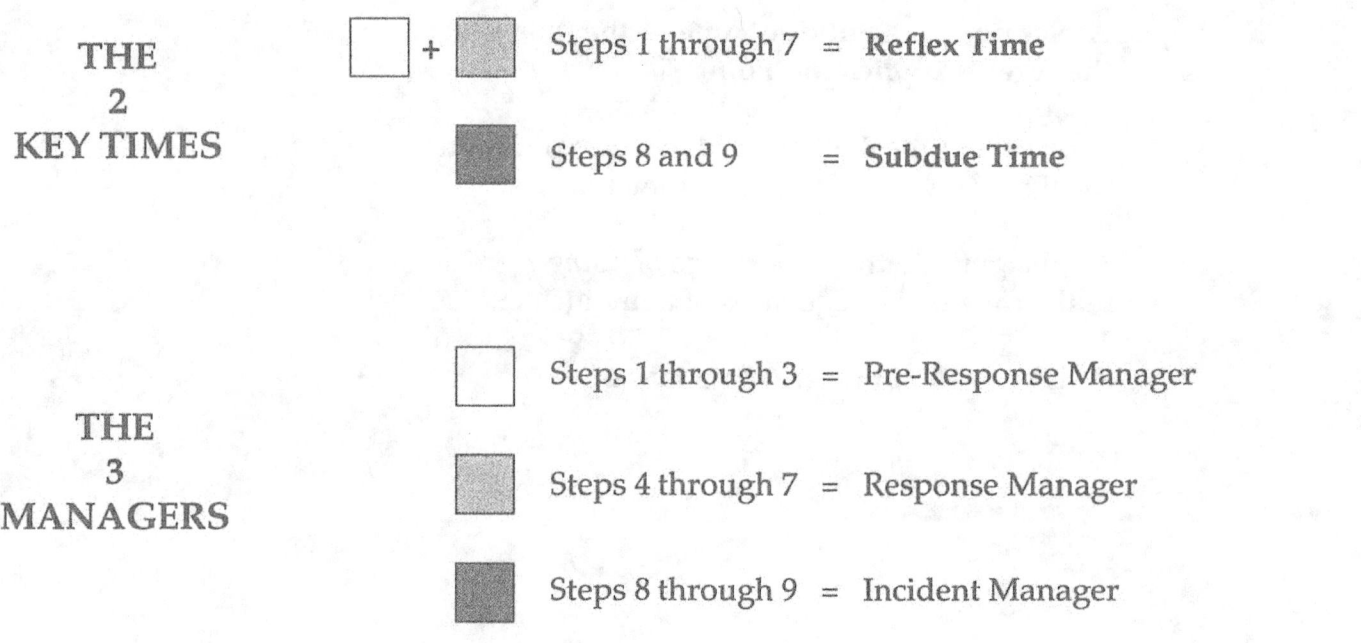

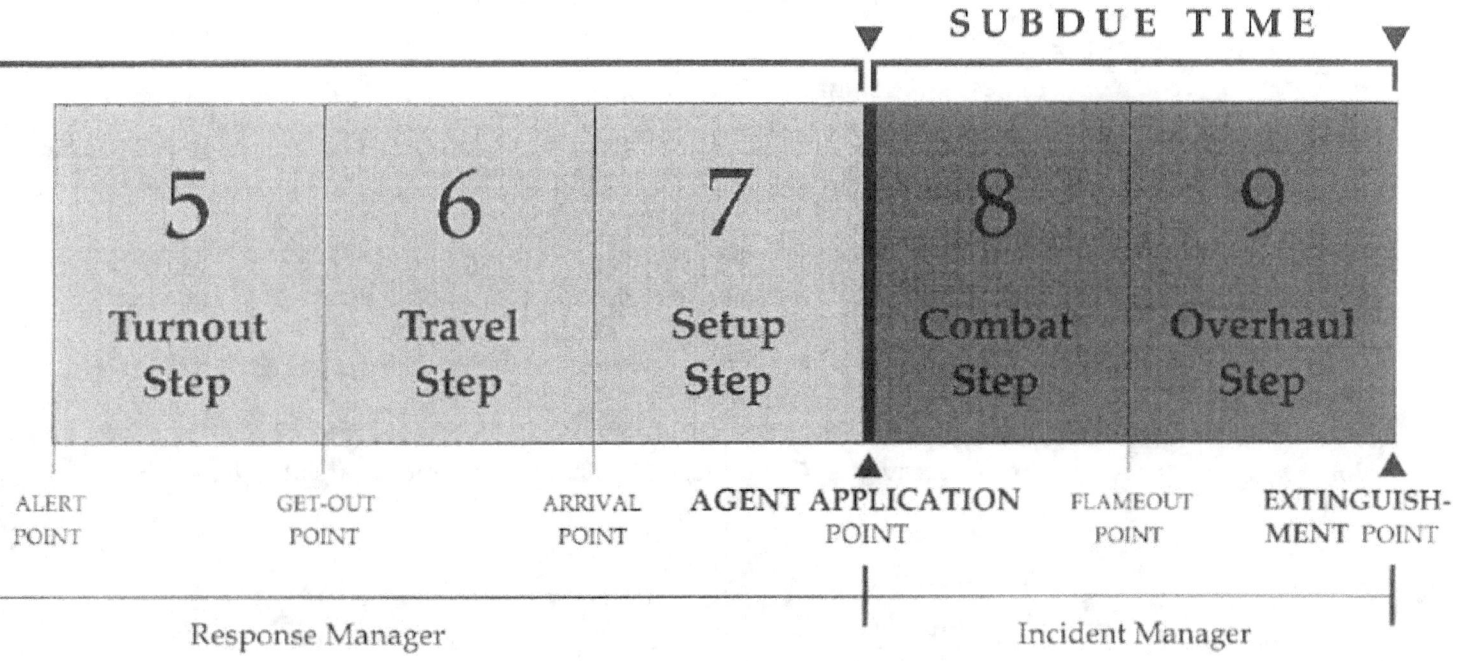

SUBDUE TIME

| 5 Turnout Step | 6 Travel Step | 7 Setup Step | 8 Combat Step | 9 Overhaul Step |

ALERT POINT — GET-OUT POINT — ARRIVAL POINT — **AGENT APPLICATION** POINT — FLAMEOUT POINT — **EXTINGUISH-MENT** POINT

Response Manager   Incident Manager

Nine Steps from Ignition to Extinguishment for Manual Fire Suppression

From the moment a fire ignites to the moment it goes out, there is a continuum of time.

When a local Fire Service must extinguish a fire, this continuum consists of nine specific time intervals or "steps." Before and after each time step, there is an identifiable "point." This document will identify each of these points and the **9 Steps** between them. The most important point is the "magic moment" of the **Agent Application Point.**

Fire Service extinguishment is generally a series operation. While rescue, salvage and overhaul operations take place during **Setup** and **Combat**, the two major steps take priority. Thus, each major step is usually completed before the next is begun. Time gained in one step is time gained in each step that follows.

Reflex Time extends from the **Ignition Point** to the **Agent Application Point**. Two managers have a part to play in reducing **Reflex Time**.

Chapter 2 defines the **9 Steps** between each two points. Chapter 3 defines **Reflex Time** and **Subdue Time**. Chapter 4 identifies the managers usually responsible for the time steps. Chapter 5 breaks down the **Combat Step** into three substeps that can be used. Chapter 6 discusses various timekeeping techniques. Chapter 7 gives a summary of this document.

WHAT IS ONE SECOND WORTH?

One second of Human Movement operates on a different time scale than one second of Fire Growth.

Human Movement

Each of us deals daily with the fact that we move at a fixed speed. We can run at 1.5 to 3 metres (5 to 10 feet) per second, or walk at 0.6 to 1.2 metres (2 to 4 feet) per second. Once on a course at a given speed, humans move linearly; that is, the distance covered in 3 seconds is the same at the beginning or later (before fatigue). In Figure 1-1 we see two equal time periods: A and B. Time period A and time period B both result in the same linear distance traveled (C_1 and C_2). Human movement occurs along a straight line; it is linear movement.

Fire Growth

While Fire Service operations must be linear, flames inside a building grow differently. Fire develops on a geometric scale; that is, it gets worse and worse as it grows. Technically, this is "exponential" growth rather than linear movement.

In Figure 1-2, the amount of damage during time period A, when the fire is just beginning, is slight (D_1). The amount of damage during time period B, however, is substantial (D_2). Suppose during B a transition point (critical point) is passed. If so, we might go from *damage* to *major damage*, or we could go from *life* to *death*.

EXAMPLE:

In a Fire Service on the West Coast (USA), all 29 stations hear the initial call for help. The dispatcher can cut off the station speakers if the message is garbled or if it's a wrong number. If the first-due engine team, first-due ladder team, and first-due chief understand the message, they may respond.

One evening during dinner at the main station, a call came in from a frantic woman nearby who reported, "Kids set fire to the outside of my house!" This station was the first-due station. Twenty-five seconds after the caller's message started, blue diesel smoke filled the now-empty apparatus floor.

On arrival, Fire Service personnel found that the fire had burned through the exterior siding and into the stud space of a two-story, balloon-constructed, wood-frame dwelling. The first-due ladder team cut open the outside wall, and the first-due engine team

Figure 1-1.
Human Movement is Linear.

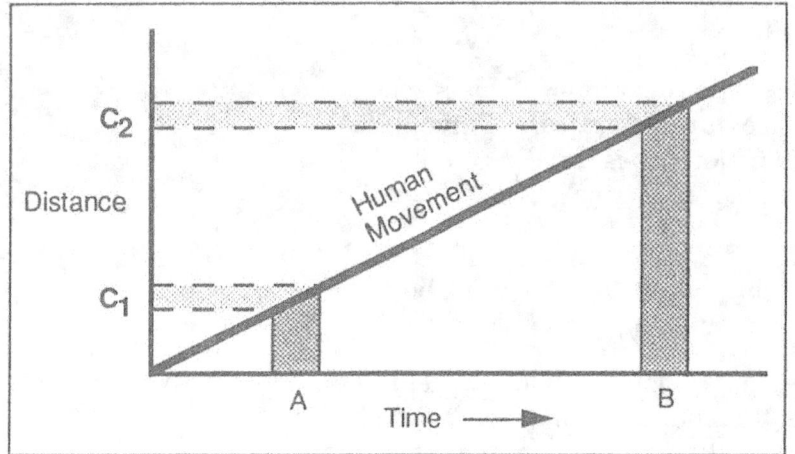

quashed the flames. The flame char was measured 1.4 metres (4-1/2 feet) up inside the wall stud space.

This observer estimated that the flames in that stud space had been 3 to 15 seconds from reaching the attic. Traditional alarm processing techniques would have added 55 to 65 seconds to the "response time" (as defined here). Had this added time been a factor, the flames would have entered the attic, and a second alarm effort would have been required. An extra 60 seconds of response time would have increased the resulting damage from about US $150, with the ability to sleep there that night — to over US $40,000, with loss of personal treasures and no sleep there for months.

Time Available

Along the bottom axis of both Figure 1-1 and Figure 1-2, two identical periods (A and B) are measured. They can be any equal periods of time — both could be one second or 60 seconds, or longer. For this discussion, let's think of 3 seconds for period A and 3 seconds for period B.

Linear Movement

Let's now think of a person moving at maximum speed. This person's progress can be shown along a sloping straight line (Figure 1-1). At a steady pace, the distance traveled in 3 seconds will be the same during A as the distance traveled during B. The resulting distances, shown as C_1 and C_2, are equal.

Exponential Growth

Now let's look at fire. Fire does not grow on a straight line; it grows on a curve (Figure 1-2). This curve is exponential. Different fuels burn at different speeds. Flames across the surface of a pool of gasoline are faster than flames on corrugated cardboard. Flames on corrugated cardboard are faster than flames on solid wood timbers. For any specific fuel bed, this growth is fixed and is determined by the particular fuel and the conditions at hand.

But 3 seconds at the beginning (A) do not produce the same effect as 3 seconds later on (B). The amount of damage to people, property, or mission is not the same. Damage is much worse the later it occurs in a fire's growth. Thus the amount of damage during time period A is only a portion of the damage during time period B. The longer an attack takes, the worse the damage gets. One second longer can be significant.

The Critical Moment

Should a critical moment (e.g., flame reaching the ceiling) be passed, the results can be as dramatic as in the example given earlier. As one East

Figure 1-2.
*Fire Growth
is Exponential.*

Coast fire chief recently put it: "It is true that 10 seconds saved at the beginning of the process (through improved dispatching or initial attack procedures) will result in agent application 10 seconds earlier, but the real issue is extinguishment. That same 10 seconds could mean the difference of hundreds of seconds, thousands of litres (gallons) of water, and dozens of personnel in the extinguishment effort."

What is One Second Worth?

One second can make a major difference in the outcome of a fire. Intelligent time-saving during manual extinguishment operations can make a dramatic difference in the resulting damage.

THOUGHTS ABOUT THE TERM "RESPONSE"

"**Response**" is a term used to gauge the reaction by the entire Fire Service organization, from the Alarm Center to the person on the nozzle. Thus, "response" is the interval between the first indication of a fire in the Alarm Center (**Alarm Point**) and the first application of extinguishing agent on the flame (**Agent Application Point**).

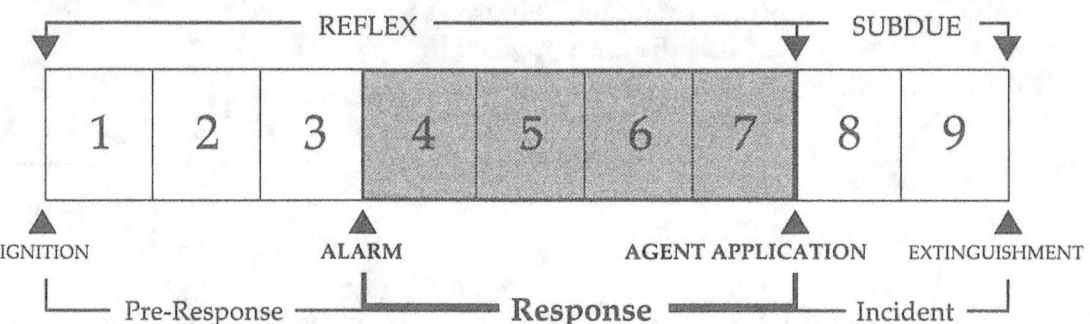

Response by a Fire Service is dictated by its established organizational policies and practices. The manager of Fire Service response is discussed in Chapter 4. Changes in policies or practices can improve response.

The station officer or the apparatus officer (managers of the **Turnout Step** and **Travel Step**) might focus only on one portion of the total departmental effort. Thus, their use of the term "response" may be limited.

In other words, the apparatus officer may use verbal shorthand by saying "response time" when actually meaning the interval of the **Travel Step** — that is, the time between the unit's departure from the station (the **Get-Out Point**) and its arrival at the scene (the **Arrival Point**).

The person collecting data must carefully determine what the person reporting the time actually means by the term "response." Is it departmental "response" or apparatus travel?

Chapter 2
Definition of the 9 Steps

This chapter will give the reader:

- A definition for each of the 9 Steps.

- A brief discussion of each Step.

- The definition of the specific Point at the start and at the end of each Step.

    Note: Each Point used in
    this chapter is "SMART".....

    S pecific
    M eaningful
    A cquirable
    R ealistic
    T imeable

"Knowledge advances by steps, not by leaps." — Baron Macaulay, 1828

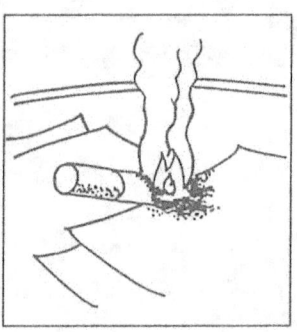

Ignition Point

The **Ignition** is at a specific point in time. *Ignition* is the first appearance of flame, whereas *burning* is the start of self-sustained non-flaming combustion in the target fuel bed. The first burning (smoldering) is a separate point from the initial **Ignition Point** (the first appearance of flame).

If a cigarette is dropped into an overstuffed cotton chair, it may cause an "over-heat" condition or "smolder" (burning). However, if the chair fabric or stuffing starts flaming, then the **Ignition Point** has occurred.

Step 1

FREE BURN STEP

The "Free Burn Step" is the period of time the fire grows before human "recognition."

The **Free Burn Step** is defined as the interval between the **Ignition Point** (first flame) and the **Recognition Point** (first human awareness). This **Free Burn Step** may be just a few seconds or many hours in length.

Recognition Point

The **Recognition Point** is the moment when the first human being senses the existence of an abnormal condition. The person may first recognize something is not right but has not yet identified it as a threat. The human being is the only mechanism that plays a role in the **Recognition Point**.

The **Recognition Point**, by definition, must always be physically sensed by a human being. Any physical instrument, whether a monitoring, warning, or detecting device, has a planned and predetermined "signal." The planned "signal" of a device is the **Detection Point**, not the **Recognition Point**.

FIREPRO₀ Institute Ltd.

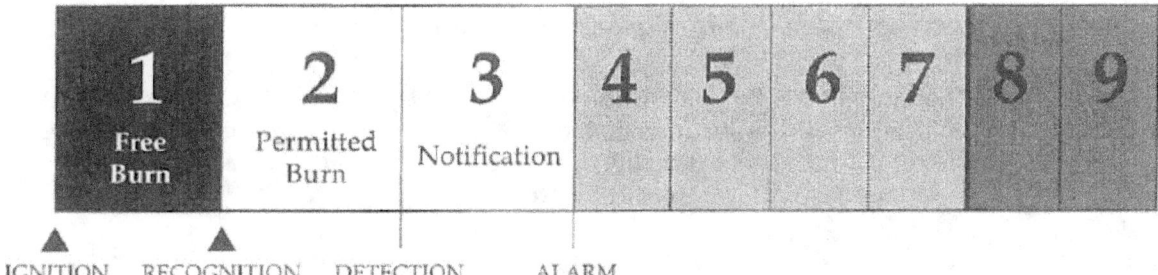

Discussion

EXAMPLE: A shopping center at 6:30 a.m.

> Present are three security guards and five employees of one anchor store who had just arrived. A janitor's closet off a side corridor at the far end of the facility from the occupied areas was the site of the first flame or the **Ignition Point**. The flames started growing, feeding on paper towels and toilet tissue stored on wooden shelving.
>
> Without automatic sprinklers or detection, the flames have to grow to produce enough smoke to travel and cause a smell to be noticed by one of the eight noses present. When the first individual becomes aware that something is not right, even if she can not be certain it is smoke, the **Recognition Point** is reached. Most noses sense smoke at 2 to 3 parts per million, well before any visible smoke is present.

During this Step, the fire may have a chance to get a foothold and consume a fair amount of combustibles. The flaming is free-burning and burning unchallenged. Thus the **Free Burn Step** is Step 1.

A similar situation can occur in a *defend-in-place* occupancy such as a hospital. Of course, more noses are usually present 24 hours a day in a hospital. The **Recognition Point** may be earlier and the flames smaller. In a *ring-and-run* occupancy, such as a school or a warehouse, at times there are fewer noses present. Thus in ring-and-run occupancies, the **Recognition Point** may be delayed and the flames larger when finally recognized.

The **Free Burn Step** may be of short duration in a fast-developing fire, while it can be a longer interval with a slowly developing fire.

If a human recognizes something abnormal *before* a device, then the **Recognition Point** applies. If a device signals the alarm first, the **Detection Point** that follows has been reached.

Recognition Point   *[repeated from previous page for convenience]*

The **Recognition Point** is the moment when the first human being senses the existence of an abnormal condition. The person may first recognize something is not right but has not yet identified it as a threat. The human being is the only mechanism that plays a role in the **Recognition Point**.

The **Recognition Point**, by definition, must always be physically sensed by a human being. Any physical instrument, whether a monitoring, warning, or detecting device, has a planned and predetermined "signal." The planned "signal" of a device is the **Detection Point**, not the **Recognition Point**.

Step 2

PERMITTED BURN STEP

The "Permitted Burn Step" is the period of time the fire grows before it is "discovered" *and* considered a threat.

The **Permitted Burn Step** is defined as the interval between the **Recognition Point** (first awareness) and the **Detection Point** (the first threat). The **Permitted Burn Step** may be a few seconds or many minutes in length.

Detection Point

The **Detection Point** is the instant when human beings first take action on sensing a definite threat, or when an automatic device first closes its contacts or releases its fusible link. The moment a person turns to leave, or the contacts close on a device, is the instant of detection. The first occurrence of such action is the **Detection Point**.

The rate of initial flame growth sometimes can be surprisingly fast. A person may see a flame that is small (**Recognition Point**) and not feel threatened as they go for a portable fire extinguisher or a pan of water. When s/he returns, flames may be large and the observer takes other action (**Detection Point**).

FIREPRO® Institute Ltd.

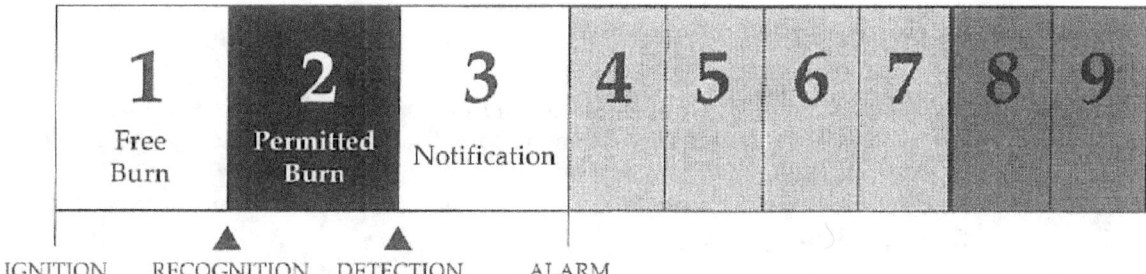

IGNITION   RECOGNITION   DETECTION   ALARM

Discussion

EXAMPLE:

A Fire Service responded to an office building at quitting time. The receptionist reported that the occupants had smelled smoke (**Recognition Point**) all day from the moment they first came in, but they had been unable to find anything wrong.

Because they didn't want to leave the building overnight with a smoke smell in it, they called the Fire Service (**Alarm Point**). Upon investigation, Fire Service personnel found a well-developed fire inside a wall of the building. The **Permitted Burn Step** spanned the working day: 28.8 kiloseconds (about 8 hours).

EXAMPLE:

Electricians had fixed the smell coming from an overheated light fixture in a shopping mall boutique. The next day, a tiny flame occurred in the concealed space above the ceiling in one of the other stores. The air circulation system allowed only a few people to be aware of the faint smell (**Recognition Point**).

Since the insulation on electrical wires was involved, the smell was similar to that of the light fixture. Some persons dismissed it as a holdover from the day before. Some persons reported it to a shop clerk and were told, "It's being taken care of." And some just didn't want to admit there was something going on.

Within an hour, the small flame had reached a new fuel source and began to grow. The smell became stronger. It soon became obvious that this was not a casual event (**Detection Point**). The shopping center authorities were notified that a fire was in progress.

Devices

In the case of a detection device, the first fusible-link release or contact closure is the **Detection Point**, rather than the **Recognition Point**.

Detection Point   *[repeated from previous page for convenience]*

The **Detection Point** is the instant when human beings first take action on sensing a definite threat, or when an automatic device first closes its contacts or releases its fusible link. The moment a person turns to leave, or the contacts close on a device, is the instant of detection. The first occurrence of such action is the **Detection Point**.

The rate of initial flame growth sometimes can be surprisingly fast. A person may see a flame that is small (**Recognition Point**) and not feel threatened as they go for a portable fire extinguisher or a pan of water. When s/he returns, flames may be large and the observer takes other action (**Detection Point**).

Step 3

NOTIFICATION STEP

The "Notification Step" is the period of time the fire grows while the alarm is being transmitted from the incident building to the local Fire Service with jurisdiction.

The **Notification Step** is defined as the interval between the **Detection Point** (the first threat) and the **Alarm Point** (the first signal received by the Alarm Center). This **Notification Step** may be a few seconds or many hours in length.

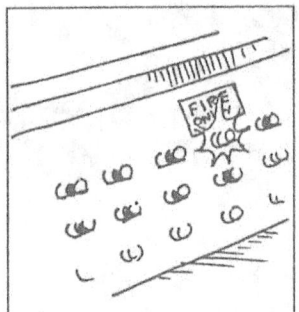

Alarm Point

The **Alarm Point** is the first sound or visual indication in the Alarm Center of the responding Fire Service. A call to 911 is not the **Alarm Point** if the call must be transferred to the Fire Service Alarm Center. When the Fire Service Alarm Center receives the call, it is the **Alarm Point**.

The most common form of the **Alarm Point** is a digit on a 911 screen or the light under the "fire only" number on a Fire Service telephone switchboard. When a 911 screen lights up, a box circuit sounder first clicks, or a telephone light first comes on, the **Notification Step** ends, and the **Alarm Processing Step** begins.

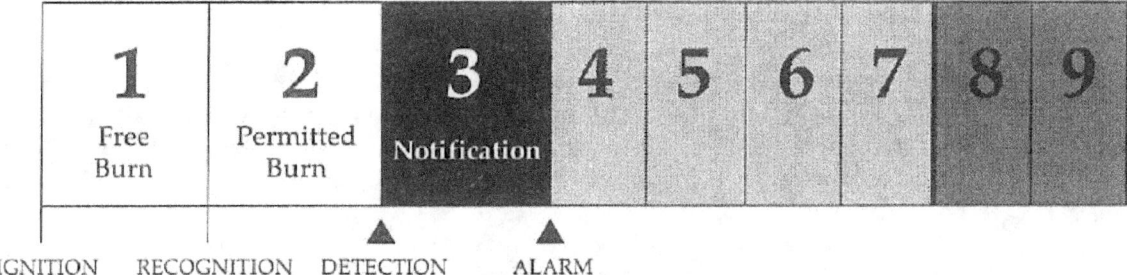

Discussion

Some installations of local detection and sprinkler equipment have a bell or gong on the outside of the building with a sign under it: "When this bell rings, call the Fire Service." This sign, of course, is addressed to "the world" rather than to a specific person. Individual people traditionally seem to ignore this sign and do not call the Fire Service. The bell or gong can ring for hours without any "alarm" arriving at the Alarm Center.

Thus, the local bell may be sounded promptly, but the **Notification Step** to the Alarm Center can be delayed for an extended time.

There are currently three methods for physically connecting a local building to an Alarm Center: "auxiliary system," "remote station system," and "central station system." Some Fire Service authorities welcome early signals, while others have "rules" and "traditions" that delay signals.

EXAMPLE:

> A sprinkler activated in a warehouse. The flow of water opened the alarm valve. The alarm valve intermediate chamber allowed water into the water-motor gong, which started its ringing. This was about 9:00 p.m. in a relatively deserted industrial area of a major city.
>
> The gong rang and an unknown number of cars passed. At about 10:30 p.m., a police car with two officers passed. More cars passed. Finally, about 3:00 a.m., a police officer who had heard the ringing gong before did radio that the gong "was still ringing." The police dispatcher called the fire dispatcher, and the response began.

It does seem a bit foolish for a business owner to spend a large amount of money for a sprinkler system or an alarm system and then not invest the small amount of dollars necessary to tie the alarm system cabinet to the local Fire Service.

The **Notification Step**, unless automatically connected to the Alarm Center, can be quite long. The degree of damage to property and mission (and, in some cases, people) resulting from this delay can be extensive.

Alarm Point   *[repeated from previous page for convenience]*

The **Alarm Point** is the first sound or visual indication in the Alarm Center of the responding Fire Service. A call to 911 is not the **Alarm Point** if the call must be transferred to the Fire Service Alarm Center. When the Fire Service Alarm Center receives the call, it is the **Alarm Point**.

The most common form of the **Alarm Point** is a digit on a 911 screen or the light under the "fire only" number on a Fire Service telephone switchboard. When a 911 screen lights up, a box circuit sounder first clicks, or a telephone light first comes on, the **Notification Step** ends, and the **Alarm Processing Step** begins.

Step 4

ALARM PROCESSING STEP

The "Alarm Processing Step" is the period of time the fire grows while the Alarm Center for that jurisdiction processes the signal to alert the responding station.

The **Alarm Processing Step** is defined as the interval between the **Alarm Point** (first light in the alarm center) and the **Alert Point** (first sound inside the responding station or an outside horn or whistle in the "town"). This **Alarm Processing Step** may be one second to several minutes in length.

Alert Point

The **Alert Point** is that moment in which the first sound occurs in or near the first responding station.

Traditionally, this is a bell, chime, tone or whistle inside a fire station, or it may be an outside air horn or siren for the "town." Usually, 4 to 15 seconds are required to transmit the sound to the point the team in the station realizes they will be responding. Some cities control the signals so that only persons in those stations responding hear the alert.

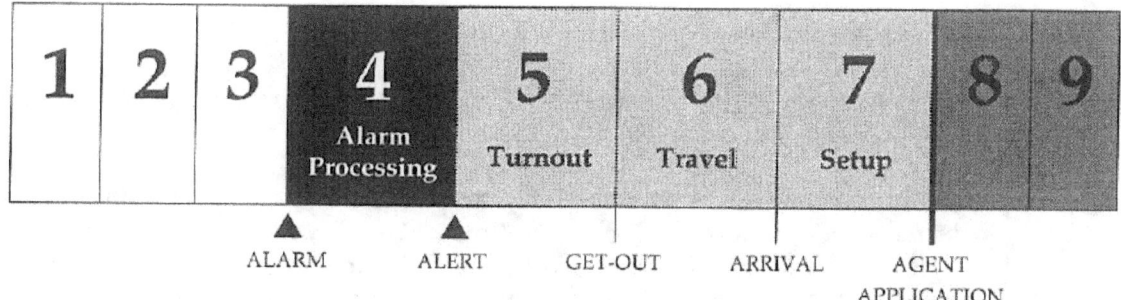

1	2	3	4	5	6	7	8	9
			Alarm Processing	Turnout	Travel	Setup		

ALARM   ALERT   GET-OUT   ARRIVAL   AGENT APPLICATION

Discussion

Promptest

One major city with 29 fire stations has a one-second **Alarm Processing Step** since the call automatically opens all fire station speakers when the alarm dispatcher answers the call. The dispatcher has a floor button to cut off speakers if the message is garbled or wrong. The first-in engine, first-in ladder, and battalion chief may respond on this verbal call if they can understand it. The rest of the alarm assignment may get ready if they wish.

Usual

In traditional receive-and-retransmit Alarm Centers, most box alarm numbers are retransmitted to the responding stations in 15 to 25 seconds. Most verbal telephone calls average 50 to 70 seconds during the **Alarm Processing Step**.

Slowest

One community's Fire Service with elaborate semi-military operations in its Alarm Center holds all fire calls "until everything is ready." This center averages 135 to 170 seconds (over 2 minutes) per call during **Alarm Processing**. Obviously, the **Alarm Processing Step** is extended during heavy call periods.

Combining Fire Alarm/Police Operators

Recent community decisions attempt to save money by combining the fire and police dispatchers. Since the vocabulary of the Fire Service is different from that of the police service and the urgency is often different, and since the operators cannot switch their vocabulary and urgency on each call, the combined dispatcher approach in busy areas usually does not work. In a recent case, a battalion chief ordering a third alarm was told to wait until the combined fire/police dispatcher would be free.

Alert Point   *[repeated from previous page for convenience]*

The **Alert Point** is that moment in which the first sound occurs in or near the first responding station.

Traditionally, this is a bell, chime, tone or whistle inside a fire station, or it may be an outside air horn or siren for the "town." Usually, 4 to 15 seconds are required to transmit the sound to the point the team in the station realizes they will be responding. Some cities control the signals so that only persons in those stations responding hear the alert.

Step 5

TURNOUT STEP

The "Turnout Step" is the period of time the fire grows while the personnel in the fire station get ready to respond.

The **Turnout Step** is defined as the interval between the **Alert Point** (first sound inside the responding station or town) and the **Get-Out Point** (front wheel crosses the threshold of the responding station). This **Turnout Step** may be as short as 15 seconds or as long as 210 seconds (3-1/2 minutes).

Get-Out Point

The **Get-Out Point** occurs when the front wheel of the responding apparatus crosses the threshold of the station. This means that the door has gone up satisfactorily and the apparatus is staffed and moving.

Volunteer departments work hard to reduce the time to the **Get-Out Point**.

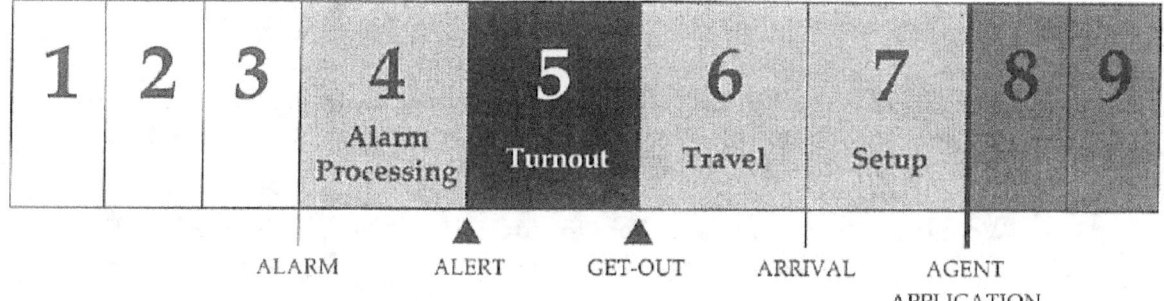

| 1 | 2 | 3 | 4 Alarm Processing | 5 Turnout | 6 Travel | 7 Setup | 8 | 9 |

ALARM   ALERT   GET-OUT   ARRIVAL   AGENT APPLICATION

Discussion

A number of important actions must be taken between the time of the alert signal in the station (**Alert Point**) and the time the apparatus leaves the station (**Get-Out Point**). This **Turnout Step** requires getting to the apparatus floor; kicking off shoes; putting on boots, coats and helmets; mounting the apparatus; starting the truck; opening the door; and getting the front tires to roll across the threshold (**Get-Out Point**).

Promptest

In the major city with the one-second **Alarm Processing Step**, the box alarm assignment is getting all their equipment ready, getting their apparatus started, opening the door and preparing to roll. If the anticipated box alarm is then sounded, the readied crew and apparatus head out the door with no delay. The **Turnout Step** is less than 5 seconds.

Slowest

Recently in a major city, a battalion chief was responding to an alarm and passed a station 120 seconds (2 minutes) into his response. The first-in apparatus was inside the station and the door was still down.

At that time, the engine company officer had the idea that he should get his crew together, brief them on what he knew about the building, look up the address and the location of the building, decide on the route that should be taken to get to the building, and then get onto the apparatus floor and get themselves ready. Once all the crew was seated properly and belted in, the officer then gave the order to open the door and start the response. The **Turnout Step** was over 180 seconds (3 minutes).

General

Rushing is not a good idea; however, placing of clothing, knowing every building in the first-in area, and following Fire Service policies can shorten the required **Turnout Step**.

Get-Out Point   *[repeated from previous page for convenience]*

The **Get-Out Point** occurs when the front wheel of the responding apparatus crosses the threshold of the station. This means that the door has gone up satisfactorily and the apparatus is staffed and moving.

Volunteer departments work hard to reduce the time to the **Get-Out Point**.

Step 6

TRAVEL STEP

The "Travel Step" is the period of time the fire grows while the apparatus is moving toward the affected building.

The **Travel Step** is defined as the interval between the **Get-Out Point** (front wheel crosses the threshold of the responding station) and the **Arrival Point** (front wheel first stops in front of the incident structure). This **Travel Step** may be a few minutes or several hours in length.

Arrival Point

The **Arrival Point** is the time when the front wheel on the first piece of response apparatus stops at the incident scene. If there are no misdirections, the building is accessible, and there are no long narrow driveways, the **Arrival Point** may put the apparatus in an effective spot to begin the attack.

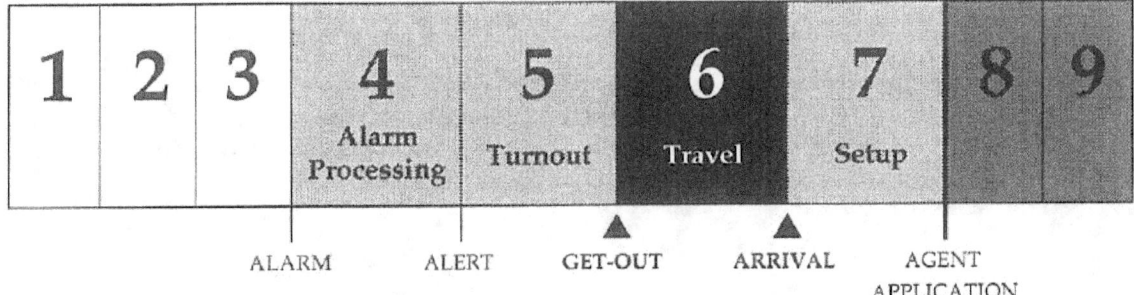

| 1 | 2 | 3 | 4 Alarm Processing | 5 Turnout | 6 Travel | 7 Setup | 8 | 9 |

ALARM   ALERT   GET-OUT   ARRIVAL   AGENT APPLICATION

Discussion

In light traffic on level ground, long pieces of heavy apparatus (such as ladder trucks) traditionally average about 180 to 210 seconds (3 to 3-1/2 minutes) per mile. Engines move at about 150 to 180 seconds (2-1/2 to 3 minutes) per mile. Light attack apparatus may average about 90 to 120 seconds (1-1/2 to 2 minutes) per mile.

Light Attack Vehicles

In one city with a large number of steep hills, the Fire Service staffs one person on a 4-wheel-drive light truck of a "Jeep" nature. This light engine carries 380 litres (100 US gallons) of water and a hard rubber hose on a reel, with a 114 litres per minute (30 gpm) spray nozzle.

While both the heavy engine and the light attack vehicle leave the station at the same time, the light attack vehicle typically arrives up to 180 seconds (3 minutes) before the heavier equipment. The flames have been burning up to 180 seconds (3 minutes) less at this time. The light attack vehicle commander has been able to extinguish some fires and hold others until the engine company arrives.

Impediments

In one community, the largest employer in town has a freight train that cuts the town in half for 7.2 to 10.8 kiloseconds (2 to 3 hours) a day, four times a week.

Thus the sum of the **Free Burn Step, Permitted Burn Step, Notification Step, Alarm Processing Step** and **Turnout Step** was the same whether the train crossing was blocked or not, but the length of the **Travel Step** was extended when the train crossing was blocked.

Fire Service management got around this situation by putting another station capable of initial attack on the opposite side of this "temporary" blockage.

Time cannot be saved by speeding up the apparatus, which moves safely only so fast. Excessive speed may contribute to the problem rather than to the solution.

Arrival Point   *[repeated from previous page for convenience]*

The **Arrival Point** is the time when the front wheel on the first piece of response apparatus stops at the incident scene. If there are no misdirections, the building is accessible, and there are no long narrow driveways, the **Arrival Point** may put the apparatus in an effective spot to begin the attack.

Step 7

SETUP STEP

The "Setup Step" is the period of time the fire grows while Fire Service personnel are preparing to apply extinguishing agent (water) on the flame.

The **Setup Step** is defined as the interval between the **Arrival Point** (front wheel first stops in front of the incident) and the **Agent Application Point** (first discharge of agent on the flames). This **Setup Step** may be less than a minute or up to several hours in length.

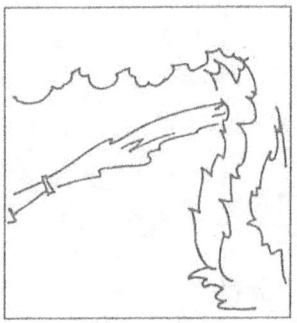

Agent Application Point

The **Agent Application Point** is the moment the first drop or particle of agent being used by the Fire Service first contacts the flame.

The **Agent Application Point** is the focus of a Fire Service and is the moment fire combat begins. The agent applied may be dry chemical from a portable fire extinguisher discharged on the flames of an automobile fire, or it may be water from a 89 mm (3-1/2 inch) preconnected blitz line discharged on the flames of a heavily involved church building.

In any case, it is the point the first Fire Service agent is discharged on the flame.

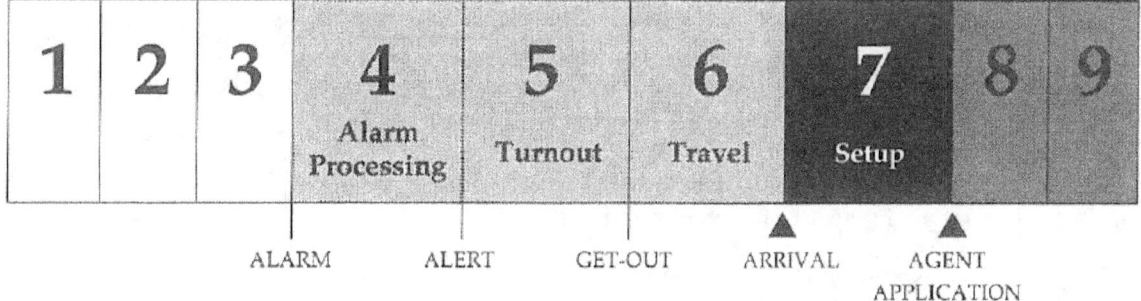

| 1 | 2 | 3 | 4 | 5 | 6 | 7 | 8 | 9 |
| | | | Alarm Processing | Turnout | Travel | Setup | | |

ALARM   ALERT   GET-OUT   ▲ ARRIVAL   ▲ AGENT APPLICATION

Discussion

The **Setup Step** requires the Fire Service officer to size up the situation and decide what is needed in order to operate. Rescue, apparatus relocation and salvage operations may occur during this period in some cases.

A number of Fire Services provide a fixed procedure for attacking large fires near the apparatus (blitz attack) in which no one has to answer the question "What is to be done?" The principle role of the officer in a blitz-trained Fire Service is to slow down the operation if it's a "small fire" on arrival, or figure out what to do if it's a different type of incident.

EXAMPLE:

One Fire Service training officer became aware that an old-time battalion chief was making his crew disconnect the preconnected hoses and re-lay engine hosebeds the way it was done when he started in the Fire Service. The training officer asked the crew on each shift to run a specific evolution while he timed the difference.

The evolution was at the training center and near the engines. The preconnected lines would reach, thus no extension of lines was required. All crews used their engines, their equipment, and the same simple evolution. The crews with preconnected lines were 30 to 45 seconds faster than crews without preconnected lines.

Some authorities are now beginning to subdivide the Setup Step into the Access Substep (getting into the building) and the Penetration Substep (getting the hoseline to the room of origin after getting in the building).

Standpipes

Once the site of the flame is more than about 46 metres (150 feet) from the apparatus position, remote hoselines or extensions will be needed. Upper floors have vertical standpipes to provide water on the floor below for fighting the fire on the floor above.

Certain Fire Services are requiring horizontal standpipes for areas beyond 46 metres (150 feet) from the apparatus at the curb. The purpose of horizontal standpipes is the same as for vertical standpipes: to shorten the **Setup Step**.

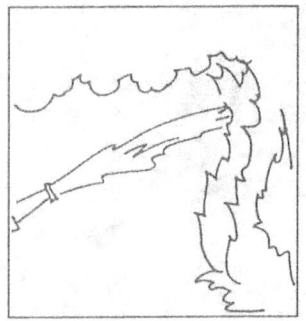

Agent Application Point   *[repeated from previous page for convenience]*

The **Agent Application Point** is the moment the first drop or particle of agent being used by the Fire Service first contacts the flame.

The **Agent Application Point** is the focus of a Fire Service and is the moment fire combat begins. The agent applied may be dry chemical from a portable fire extinguisher discharged on the flames of an automobile fire, or it may be water from a 89 mm (3-1/2 inch) preconnected blitz line discharged on the flames of a heavily involved church building.

In any case, it is the point the first Fire Service agent is discharged on the flame.

Step 8

COMBAT STEP

The "Combat Step" is the period of time the flame is continuously visible while agent is being discharged on the flames.

The **Combat Step** is defined as the interval between the **Agent Application Point** (first discharge of agent on the flames) and the **Flameout Point** (all flame disappears for the last time). This **Combat Step** may be a few seconds or many days in length.

Flameout Point

The **Flameout Point** is the first moment all flaming ceases. There may be the usual flames during **Overhaul** after the **Flameout Point**.

If the fire has not reached Full Room Involvement at the **Agent Application Point** for normal-sized rooms, **Flameout** generally can be accomplished in less than a minute (a Quash). The larger the room, the more skill is needed to reach the **Flameout Point**. [Quash is discussed in Chapter 5.]

Once the fire has reached 280 square metres (3,000 square feet) in size, any Fire Service will be on the defensive and must wait for the finer fuels to be consumed, so they can move in and finally cause the **Flameout Point** to occur.

FIREPRO® Institute Ltd.

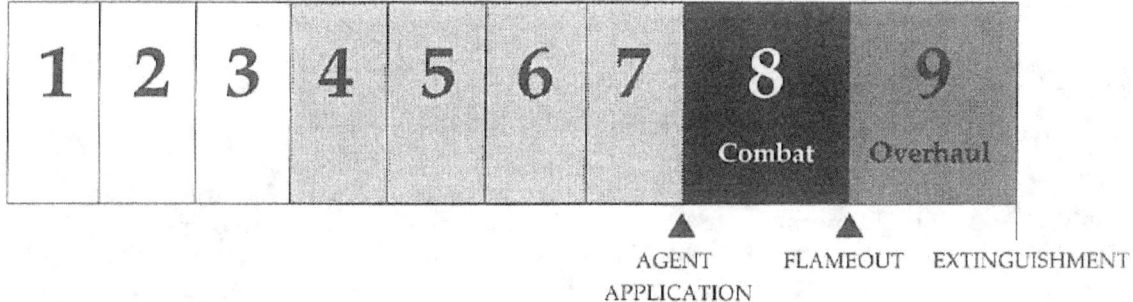

Discussion

EXAMPLE:

Fire Service personnel responded to a dwelling fire to find the living room fully involved and flame coming out a window. A 38 mm (1-1/2 inch) hoseline with a fog nozzle discharged through the window for less than 30 seconds produced a flame-free environment with some smoldering. The **Combat Step** was less than a minute.

EXAMPLE:

In the MGM fire in Las Vegas, the Fire Service responded and found itself immediately in a defensive position. Firefighters were forced to stay "outside" for over an hour until the 11 large ventilation fans could be shut down. Once the giant fans were shut down, an advance could be made to complete extinguishment of the flame and to rescue the terrified occupants.

Leave No Water

The London Fire Service started it, and in the Seattle Fire Service certain battalion chiefs are practicing it: The attempt is to extinguish fire with no water left on the floor — use just the right amount of water to reach the **Flameout Point** and no more. Only skilled fire officers attempt this.

Generally, the larger the area involved in flame upon arrival, the longer it takes to achieve the **Flameout Point**. When the fire is above 280 square metres (3,000 square feet) of involvement, the tactic for any Fire Service is generally defensive. A defensive attack is to surround the involved area, protect the barriers, and wait for the "fine fuel" to burn so that advances can be made on the less fierce, "heavier fuels."

Rescue, salvage and overhaul operations are used where needed during this period.

Flameout Point   *[repeated from previous page for convenience]*

The **Flameout Point** is the first moment all flaming ceases. There may be the usual flames during **Overhaul** after the **Flameout Point**.

If the fire has not reached Full Room Involvement at the **Agent Application Point** for normal-sized rooms, **Flameout** generally can be accomplished in less than a minute (a Quash). The larger the room, the more skill is needed to reach the **Flameout Point**. [Quash is discussed in Chapter 5.]

Once the fire has reached 280 square metres (3,000 square feet) in size, any Fire Service will be on the defensive and must wait for the finer fuels to be consumed, so they can move in and finally cause the **Flameout Point** to occur.

Step 9

OVERHAUL STEP

The "Overhaul Step" is the period of time flame no longer burns but before all smoldering or glowing embers are extinguished.

The **Overhaul Step** is defined as the interval between the **Flameout Point** (all flame disappears for the last time) and the **Extinguishment Point** (all glowing and smoldering has ended for the last time). This **Overhaul Step** may be as short as a few minutes or many days in length.

Extinguishment Point

The **Extinguishment Point** occurs when all smoldering or glowing ceases. This **Extinguishment Point** indicates that conditions have returned to the fire-free level. The landscape may have changed, but the direct threat has ended.

Sometimes it is possible to believe that all smoldering has been quenched when, in fact, it has not. The embarrassing "rekindle" may then occur. A new alarm to a rekindle is a restart of the entire timing process for a continuation of the same fire.

On a fire that is small on arrival, the **Agent Application Point** and **Extinguishment Point** may be as short as 15 seconds apart. In a landfill area, time between the **Agent Application Point** and the **Extinguishment Point** may be over a week.

FIREPRO® *Institute Ltd.*

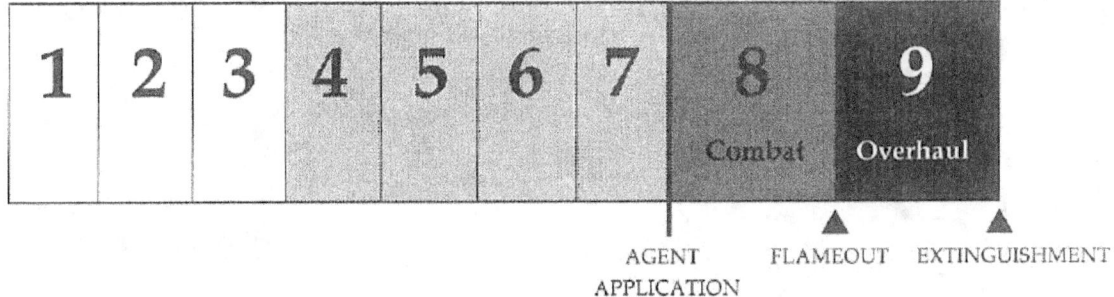

Discussion

As a general rule, fires in buildings require that they be extinguished completely. This means digging out the remains and seeing that water gets onto each of the remaining glowing areas. During the **Overhaul Step**, flareups with flame can be expected after the **Flameout Point**.

The **Overhaul Step**, or "Mop-Up" as it is sometimes called, may be an extremely difficult and time-consuming part of the operation if a large area has been damaged by the fire. Overhaul and salvage efforts may well be taking place during the **Combat Step**.

Generally, the shorter the time to **Flameout**, the easier the **Overhaul Step**.

EXAMPLE:

> In one case, the snow around an industrial plant was two feet deep. There were no other buildings within 300 metres (about 1,000 feet). The plant was totally destroyed in the fire, with all the walls and ceilings and roofs down.

> Fire Services from many surrounding communities picked up and went home, leaving the smoldering remains since the building was already destroyed. The remains would not cause any rekindle, the deep snow prevented extension, and the surrounding area had no buildings. The remains smoked for at least three days.

While buildings can cause difficult **Overhaul Step** operations, landfill areas can give smoldering conditions (overhaul) that sometimes last several weeks.

Again, larger fires generally result in longer **Overhaul Steps**. Flames are out, but ending all smoldering may take a considerable period of time and extensive personnel effort.

Chapter 3
Combining the Steps:
2 Key Times

This Chapter will give the reader:

- The definition of **Reflex Time**.

- The 7 Steps that make up **Reflex Time**.

- The "magic moment": **Agent Application Point**.

- The definition of **Subdue Time**.

- The 2 Steps that make up **Subdue Time**.

The Fire Service's basic purpose is to remove people from harm's way and put the fire out.

The two major Fire Service efforts are: to get agent on the flames — Reflex Time, and then to use those agents to extinguish flames and smoldering — Subdue Time.

"Science is organized knowledge." — Herbert Spencer

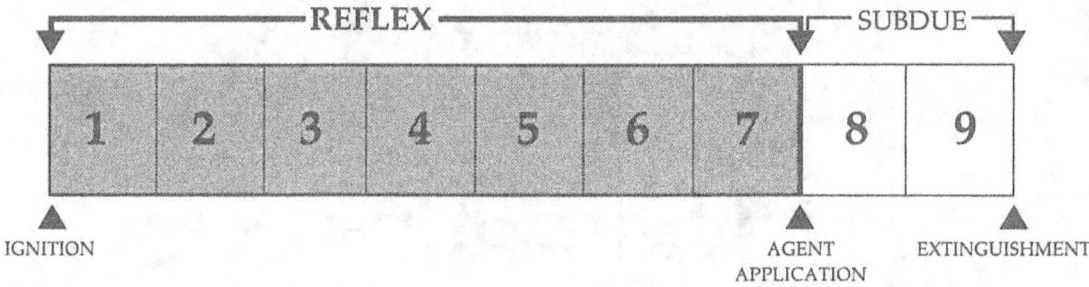

REFLEX TIME

Reflex Time is the interval measured from the **Ignition Point** to that most important **Agent Application Point** (the "magic moment").

Reflex Time is a series operation that is the sum of 7 Steps: the **Free Burn Step** plus the **Permitted Burn Step** plus the **Notification Step** plus the **Alarm Processing Step** plus the **Turnout Step** plus the **Travel Step** plus the **Setup Step**.

It is obvious that **Reflex Time** includes those items before the **Alarm Point** which are controlled by the property manager, as well as the time after the **Alarm Point** which the Fire Service manager controls.

The role of these two managers in controlling **Reflex Time** is discussed in the next chapter.

EXAMPLE:

> How much fire can a Fire Service extinguish — not just hold, but extinguish? To find out, Fire Service operations in a small city were reviewed and analyzed by a committee of chief officers of that Fire Service. They then "sold" their findings to the city council. (How they convinced the council is another story.) The result was that city government required all buildings over 930 square metres (10,000 square feet) in size to be protected with automatic sprinklers, with alarm signals directly connected to the city's Alarm Center.

> All buildings 465 to 938 square metres (5,000 to 10,000 square feet) in size were outfitted with automatic alarm devices, again with alarm signals directly connected to the Alarm Center.

This example is a well-managed use of presently available equipment to reduce **Reflex Time** in a practical way, before the ignition of any fire.

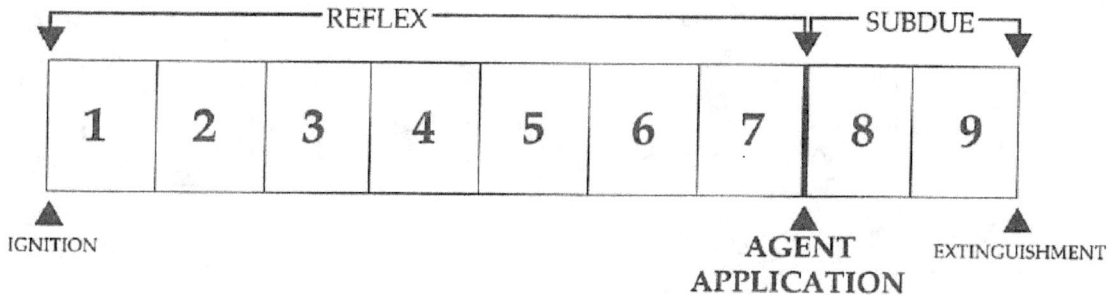

AGENT APPLICATION POINT

The **Agent Application Point** is the first moment the first drop or particle of agent being used by the Fire Service first contacts the flame. It has been called the "magic moment" of manual fire suppression.

The **Agent Application Point** is a most important point for a Fire Service and is the moment fire combat begins. The agent applied may be dry chemical from a portable fire extinguisher discharged on the flames of an automobile fire, or it may be water from a 89 mm (3-1/2 inch) preconnected blitz line discharged on the flames of a heavily involved church building.

EXAMPLE:

A county Fire Service adopted the blitz attack approach. A pre-connected 89 mm (3-1/2 inch) hoseline with a nozzle rated at 1893 liters per minute (500 gpm) was installed on each engine. Personnel were trained in its use.

Three weeks later, a first-due engine responding to an alarm arrived at a church just as the sanctuary became fully involved. The blitz line was held by three persons at the front door. The door was cracked open and the nozzle discharged. The flames were out in less than 60 seconds (1 minute).

The insurance agent was furious. He had to pay for cleaning up the 6 mm (1/4-inch) char on all the exposed wood and for the restoration. The church still has services in its building.

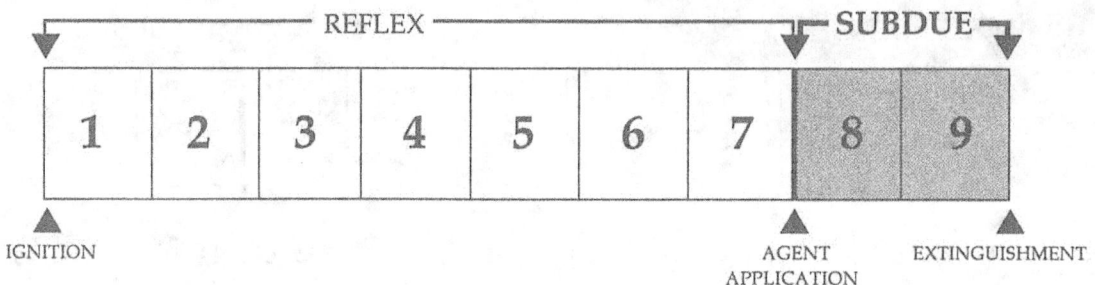

SUBDUE TIME

Subdue Time is the interval measured from the **Agent Application Point** to the **Extinguishment Point**. It is the sum of the **Combat Step** plus the **Overhaul Step**.

Subdue Time is that portion of the fire sequence controlled by the Incident Commander on the fireground.

Some non-flaming fires (woodchip piles, landfill and the like) may take skill and effort, it is true; but it is flaming fires inside buildings, structures, or vehicles that most generally are fast and threaten persons.

EXAMPLE:

A Fire Service arrives to find the living room of a fully involved dwelling. Flame is coming out of one living room window. The Fire Service attacks with water from a preconnected 38 mm (1-1/2 inch) hoseline. The nozzleman discharges this hoseline into the window for 30 seconds with a spray pattern and completely knocks down the fire and all smoldering and glowing.

The moment the first drop of water hits the flame, **Reflex Time** ends and **Subdue Time** starts. The **Subdue Time** was about 30 seconds.

EXAMPLE:

The fire in Windsor Castle grew rather rapidly on construction materials to involve a number of rooms before the arrival of the first Fire Service apparatus. The setup of initial combat hoselines began.

Supplemental apparatus was called, and additional hoselines were set until the fire in one part of the castle was surrounded. Then the Fire Service had to wait until the finer fuels — kindling fuels — burned out.

At this point, Fire Service personnel were able to move in and extinguish the flame. **Overhaul** took some time, as there were a great number of smoldering piles. The **Subdue Time** was thus many hours in length.

FIREPRO Institute Ltd.

Chapter 4

Combining the Steps:
3 Different "Managers"

This chapter will give the reader:

- A view of the role of the Property Owner's management responsibility.

- A view of the role of the Fire Service Chief Officer's management responsibility.

- A view of the role of the Fire Service Incident Commander's management responsibility.

"You earn leadership by practicing management." — Anonymous

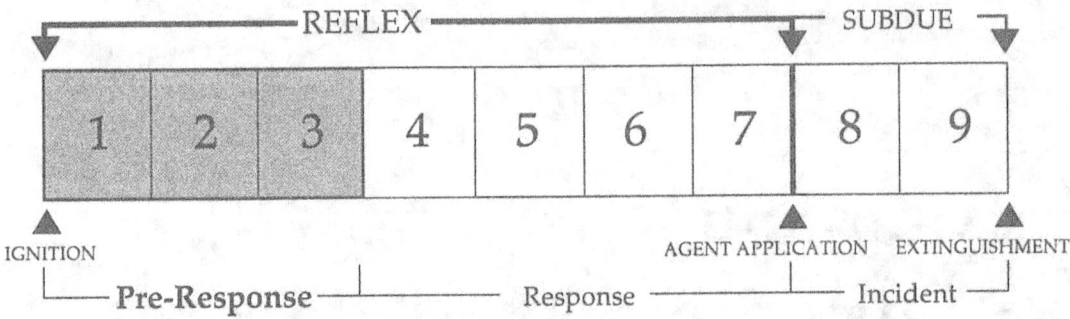

PRE-RESPONSE MANAGER

The Property Owner

The Pre-Response Segment is that portion of the **Reflex Time** controlled by the property owner. The Pre-Response Segment runs from the **Ignition Point** to the **Alarm Point**. The Pre-Response Segment is the sum of: **Free Burn Step** plus **Permitted Burn Step** plus **Notification Step**.

The largest delay in the **Reflex Time** usually occurs in the Pre-Response Segment. In a building with no automatic detection or fire sprinklers, *and* no extension of that alarm to the Alarm Center, *and* no occupants, the fire must be seen by an outsider. Therefore, the flames can and do grow to a fairly large size before they are seen. The outsider then must find a means to get the alarm to the Alarm Center, which may mean waking neighbors or driving some distance to find a phone.

The Pre-Response Segment is the responsibility of property management. The decisions property management makes on providing equipment and notification linkage to the Alarm Center can provide for smaller flames at the time that the message arrives at the Alarm Center (the **Alarm Point**).

EXAMPLE:

> A fire in the Neiman-Marcus store grew to a size that it could finally be seen outside the building. By this time, an outside observer called it a "Flash Fire."

> This passerby pounded on the front door and woke the night watchman inside. The night watchman then called for help from the local Fire Service. The Pre-Reponse Segment was longer than necessary.

Note: What is a "flash fire"? When a passerby says "flash fire," they mean fire that has already grown to a size where flames can move rapidly before they see it. A person then sees it and says "flash fire" when they actually mean, "I saw the flames move rapidly in accumulated gases — like a flash." The fire, on the other hand, was growing normally before the passerby saw it. The fire like this is not a "flash fire."

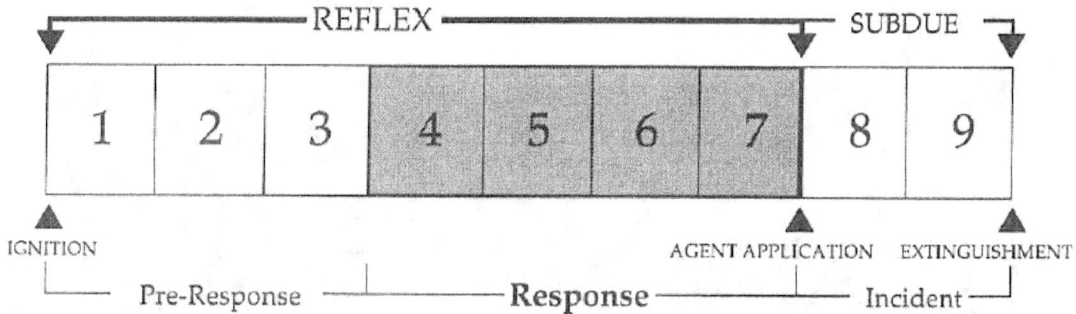

IGNITION   AGENT APPLICATION   EXTINGUISHMENT

Pre-Response   **Response**   Incident

RESPONSE MANAGER

The Chief of the Fire Service

The Fire Service's policies and procedures govern the time from the **Alarm Point** to the **Agent Application Point**. This period of time is traditionally called "Response Time." Fire Services vary in where they give "command" to the Incident Commander. The most frequently found is the **Agent Application Point**, as used in this document. Other points are valid when clearly established in a particular Fire Service.

The Response Segment is the portion of **Reflex Time** controlled by management of the Chief of the Fire Service in that jurisdiction. The Response Segment runs from the **Alarm Point** to the **Agent Application Point**. The Response Segment is the sum of: **Alarm Processing Step** plus **Turnout Step** plus **Travel Step** plus **Setup Step**.

EXAMPLE:

> The community involved had two "Mill Valleys." One, in the southern part of town, was very widely known as "Mill Valley." An old mill was situated there with three dwellings right around it. The other "Mill Valley," in the northern part of town, was called "Mill Valley" only by the five residents there.

> The call was received for a burning dwelling in "Mill Valley." The dispatcher sent the equipment to the well-known Mill Valley. The fire was actually in the lesser-known Mill Valley, seven miles away.

> It is probable that the man killed in the fire was already dead at the time the original alarm was received by the Alarm Center, but the delay of having the apparatus respond in one direction to the well-known Mill Valley instead of the other direction to the lesser-known Mill Valley extended the **Reflex Time**.

> The **Alarm Processing Step** and the **Turnout Step** were normal, but the **Travel Step** was extended.

> The longer **Travel Step** extended the **Reflex Time**.

DEFINITION:
"Response" is departmental response; that is, the interval of time from the Alarm Point to the Agent Application Point. (See page 6 for a discussion of the term "response.")

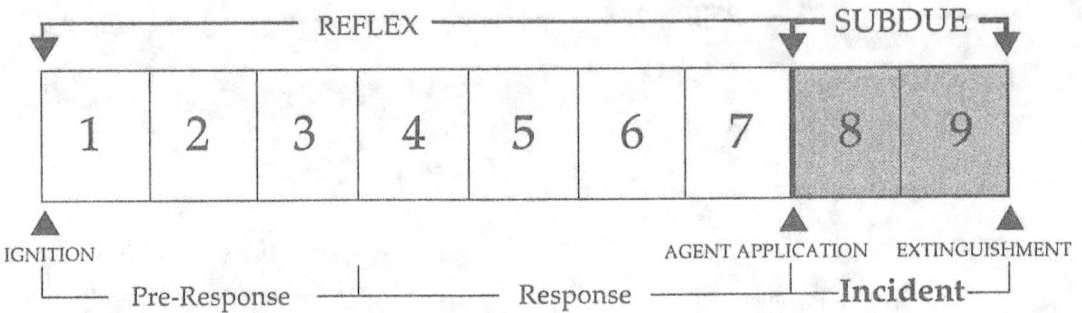

INCIDENT MANAGER

The Incident Commander

Finally it is the Incident Commander who takes charge and manages the tactical situation on the fireground. The Incident Segment runs from the **Agent Application Point** to the **Extinguishment Point** and beyond. The Incident Segment is the sum of: **Combat Step** plus **Overhaul Step**.

Before the **Agent Application Point**, the Incident Commander is responsible for submitting his ideas through appropriate channels to get policies, procedures and training mechanisms set up to shorten the time up to the **Agent Application Point**.

EXAMPLE:

> In one battalion of a large city Fire Service, the engine company officers and the ladder company officers decided to work together. The job of the ladder company officer was to open the building side or roof exactly in the area nearest the fire. The engine company officers were to use hoselines to push the flame out the hole where they thought the ladder company officer would open it.
>
> Both teams were working independently, without communication. When the hole was cut in the right place and the flame was pushed out that hole, the damage was incredibly low. The water residue was very low, and the "stop" was impressive.
>
> When the ladder company officer didn't guess the right place to open, the engine company could not push the flame out the opening, or vice versa. The resulting fire damage was no worse than it would have been without the aggressive "open and push" attack.

FIREPRO® Institute Ltd.

Chapter 5
Thoughts on Fireground Action

This chapter will give the reader:

- The **3 Fireground Functions** performed by the Fire Service:

      The Scout,
      The Quash, and
      The Operation.

- Definitions of 3 Opinion Steps within the **Combat Step**:

      Growing Substep
      Holding Substep
      Shrinking Substep

   These terms are the opinion of a skilled individual and as such they may be useful.

- Definitions of the **Containment Point** and the **Control Point**.

"Not everything that counts can be counted, and not everything that can be counted counts." — Albert Einstein

3 FIREGROUND FUNCTIONS

NOTE:
The reader may find
these terms "new."
Use the ones currently
popular in your area, or
adapt them as meaning-
ful to you.

As a Fire Service approaches and operates on the fireground, three separate functions are performed in the following order:

- The Scout Function
- The Quash Function
- The Operation Function

Scout

The Fire Service officer in the righthand seat of the first-responding apparatus has a "scouting" function. His job is to report over the radio what he sees as he approaches. This initial message is generally in the form of one of the following:

- "nothing showing"
- "smoke showing," or
- "flame showing."

Some Fire Services expand the Scout function by indicating building type and view. A typical report might be:

- "Smoke showing at the eaves of a two-story wooden dwelling. I can see two sides of the dwelling."

Quash

If the flame is knocked down in 240 seconds (4 minutes) or less from the **Agent Application Point**, the fire is said to have been "Quashed." The amount of flame that can be quashed will depend on the amount of water the initial team attacklines can discharge. 114 liters per minute (30 gallons per minute) will not Quash as much flame as 380 lpm (100 gpm); 380 lpm (100 gpm) will not Quash as much as 1893 lpm (500 gpm) can.

Operation

If "the Quash" cannot occur, then an Operation with other heavy units is required. This may be 2 vehicles or 100 vehicles, depending upon the situation and the Fire Service size.

EXAMPLE:

A chief from a neighboring department watched an operation with 61 jets (hoselines).

"I would handle that same fire with only 14 jets," he said.

"How could you do that?" he was asked.

"That's all I have" was the reply.

Offensive Operations

While the fire is relatively small and enough staff is present, the Operation can be handled safely by an inside firefighting effort. This attack is said to be "the offense." This offensive type of Operation may be successful.

Defensive Operations

Once the flame reaches a complicated stage [on two floors] or a large stage [approximately 280 square metres (3,000 square feet)], then the Fire Service is generally forced to surround the area and hold the flame inside that area. This is said to be a defensive operation. Though a lot of water is thrown at the flames, the Fire Service is kept out until the thin ends of the fuel have burned away. After the flame begins subsiding, it is possible to move in and start the **Overhaul Step**.

3 OPINION STEPS

Step 8, the **Combat Step**, can be further divided into 3 Substeps:

- The **Growing Substep**
- The **Holding Substep**
- The **Shrinking Substep**

The intervals of time in these 3 Substeps are separated by points determined by the informed and skilled opinion of the Incident Commander, rather than by an observable physical happening. A specific department or agency may have defined these intermediate points for local usage.

Definitions of these 3 Substeps, as they are being used, are given on the following pages.

THE 3 SUBSTEPS

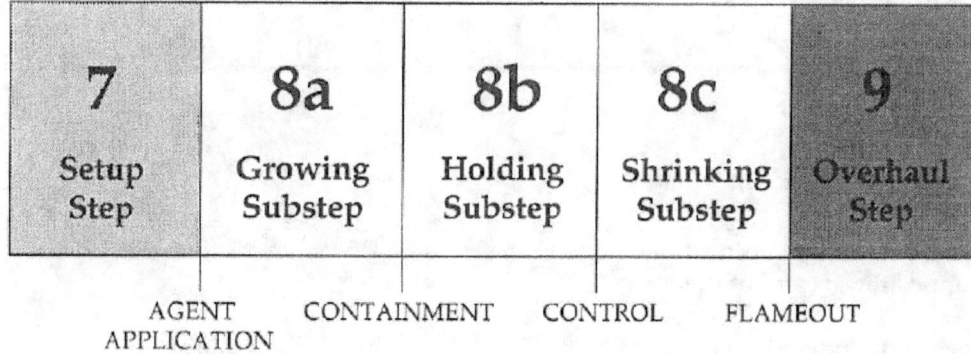

7	8a	8b	8c	9
Setup Step	Growing Substep	Holding Substep	Shrinking Substep	Overhaul Step

AGENT APPLICATION   CONTAINMENT   CONTROL   FLAMEOUT

1	2	3	4	5	6	7	8a	8b	8c	9

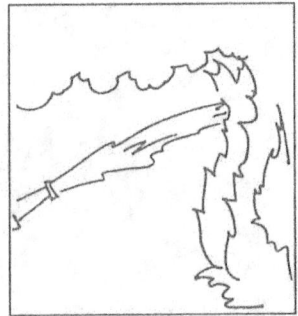

Agent Application Point *[existing, repeated for convenience from page 20]*

The **Agent Application Point** is the moment the first drop or particle of agent being used by the Fire Service first contacts the flame.

The **Agent Application Point** is the focus of a Fire Service and is the moment fire combat begins. The agent applied may be dry chemical from a portable fire extinguisher discharged on the flames of an automobile fire, or it may be water from a 89 mm (3-1/2 inch) preconnected blitz line discharged on the flames of a heavily involved church building.

In any case, it is the point the first Fire Service agent is discharged on the flame.

Step 8a

GROWING SUBSTEP

The "Growing Substep" is the period of time the flame is growing while Fire Service personnel are surrounding it.

The **Growing Substep** is defined as the interval between the **Agent Application Point** and the **Containment Point**. The **Growing Substep** is defined by the Incident Commander's opinion. The **Growing Substep** may be stopped within seconds of the Fire Service discharging water, or it may last for many hours.

Containment Point

An opinion point

The **Containment Point** is the moment the Incident Commander feels that he has stopped the growth of the fire.

The **Containment Point** may be protecting barriers inside the building or abandoning the building to the flame.

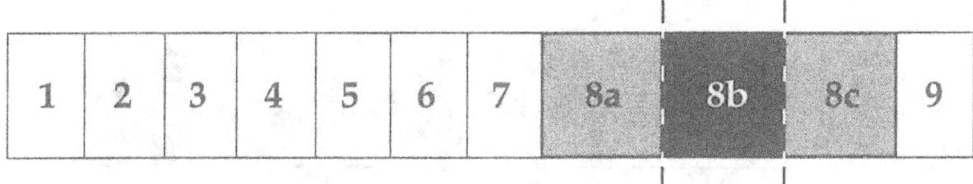

| 1 | 2 | 3 | 4 | 5 | 6 | 7 | 8a | 8b | 8c | 9 |

Containment Point   *[repeated from previous page for convenience]*

The **Containment Point** is the moment the Incident Commander feels that he has stopped the growth of the fire.

An opinion point

The **Containment Point** may be protecting barriers inside the building or abandoning the building to the flame.

Step 8b

HOLDING SUBSTEP

The "Holding Substep" is the period of time the flame continues to burn at nearly the same rate despite discharge of water on it.

The **Holding Substep** is defined as the interval between the **Containment Point** and the **Control Point**. The **Holding Substep** is defined by the Incident Commander's opinion. The **Holding Substep** may be relatively short, in terms of seconds, or it may be hours in length.

Control Point

The **Control Point** is the moment the Incident Commander feels that he has enough personnel and equipment at the scene or in the staging area to control the flame.

An opinion point

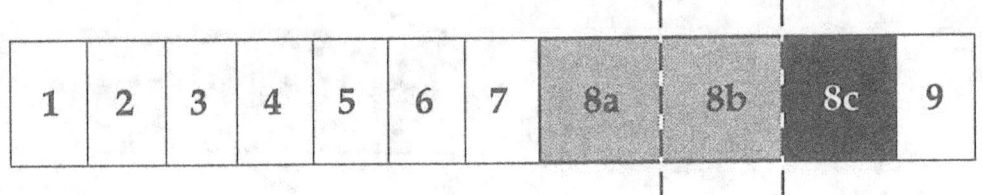

| 1 | 2 | 3 | 4 | 5 | 6 | 7 | 8a | 8b | 8c | 9 |

Control Point   *[repeated from previous page for convenience]*

An opinion point

The **Control Point** is the moment the Incident Commander feels that he has enough personnel and equipment at the scene or in the staging area to control the flame.

Step 8c

SHRINKING SUBSTEP

The "Shrinking Substep" is the period of time the flame is reducing in size as Fire Service personnel move in on it.

The **Shrinking Substep** is defined as the interval between the **Control Point** and the **Flameout Point**. The **Shrinking Substep** is defined by the Incident Commander's opinion. This **Shrinking Substep** may be minutes or hours in length.

Flameout Point   *[existing, repeated for convenience from page 22]*

The **Flameout Point** is the first moment all flaming ceases. There may be the usual flames during **Overhaul** after the **Flameout Point**.

If the fire has not reached Full Room Involvement at the **Agent Application Point** for normal-sized rooms, **Flameout** generally can be accomplished in less than a minute (a Quash). The larger the room, the more skill is needed to reach the **Flameout Point**. [Quash is discussed on page 36.]

Once the fire has reached 280 square metres (3,000 square feet) in size, any Fire Service will be on the defensive and must wait for the finer fuels to be consumed, so they can move in and finally cause the **Flameout Point** to occur.

FIREPRO® Institute Ltd.

Chapter 6
Timekeeping Action

This chapter will give the reader guidance in: *Points are timeable.*

- Collecting data elements.

- Analyzing the data collected.

- Acting on results of the analysis.

"The solution, when found, will be obvious." — Robert R. Undegraff

TIMEKEEPING

There are five "points" during fire incident time that are easy for the Fire Service to capture. Each one of these five is routinely captured by some of the Fire Service organizations around the world.

These "points" are the time of:

- receipt of the initial **Alarm**,
- **Get-Out** by the first piece of equipment,
- **Arrival** of the first piece of equipment at the scene,
- first **Agent Application** by the Fire Service, and
- "**Everyone**" back in Service.

The first three "points" are easily collectible in all fire incidents.

The fourth is an important "point" in timing of fire incidents. It shows the Fire Service manager or training officer what form of initial attack was implemented — for example, was it a portable fire extinguisher or a hoseline; was it within preconnected range or did it require hose extensions? When was the agent applied? What was the **Setup Step** timing?

The last "point" listed above, although not a point in the ignition-to-extinguishment series, is important for Fire Service management and resource allocation.

Types of Time Collection

There are two types of time collection:

- the **permanent collection** of certain time "points" on all fire incidents, and

- the **special study** of certain time "points" on all or some specific incidents.

In the *Permanent Collection* area, averages can be kept. Suppose two different Fire Services analyzed the elapsed time between the **Alarm Point** and the **Agent Application Point**. One Fire Service has an average of 453 seconds (7 minutes, 33 seconds), and the other has an average of 576 seconds (9 minutes, 16 seconds). Suppose, however, that the response distances were the same in both these agencies — why the difference in the response times?

In the *Special Study* area, a training officer of a large city Fire Service found that one shift had larger fires. A special study showed that the **Arrival Point** was the same on all three shifts, but the **Setup Step** was 55 seconds longer on the shift in question. Training could now be specifically targeted to one problem on that one shift.

TIME ANALYSIS

A Fire Service chief from the midwestern USA felt by instinct (a form of special study) that the old hard-suction connection to the fire hydrant was slow, yet the hard-suction connection was vigorously supported by a highly respected deputy. The chief created a Water Supply Study Committee and appointed the deputy as chair. The committee's charge was to cut 30 seconds off the water supply operation. It had taken a 4-person team 286 seconds (just under 5 minutes) to do it the old way, a total of 1144 team-seconds.

The Committee reported that a soft-suction connection from the side of the apparatus could be done with 2 people in 74 seconds, a total of 148 team-seconds. They had saved 996 team-seconds.

Another Fire Service found that a front-mounted 6-metre (20-foot) soft-suction connection allowed the driver to position the engine better and cut elapsed time to 1 person in 47 seconds, saving another 101 team-seconds.

Still another Fire Service found that a front-mounted 7.6-metre (25-foot) soft-suction connection did not lengthen the time used, but it did get rid of the kinks (reduced water volume) they had been experiencing.

Using Time to Your Advantage

Suppose you know that, within your Fire Service, Engine 2 and Engine 7 have the same running distance. When you start routine collection of time from **Alarm Point** to **Agent Application Point**, you find that Engine 2 has an average time of 424 seconds (7 minutes, 4 seconds) while Engine 7 has an average time of 538 seconds (8 minutes, 58 seconds). What do you do?

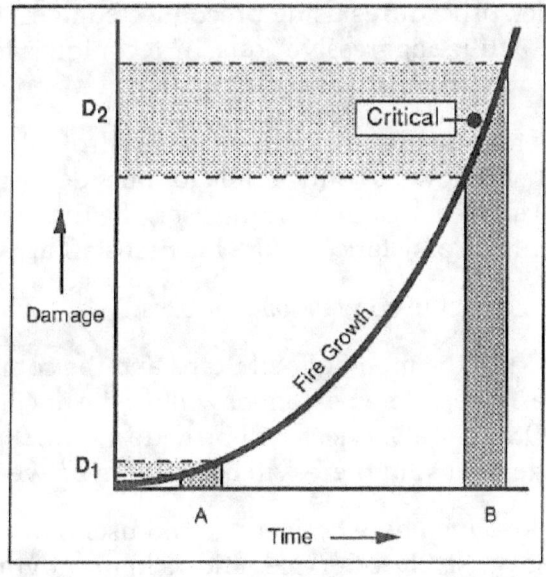

Figure 6-1.
*Fire Growth
is Exponential.*

Time analysis can be your ally in improving delivery of service.

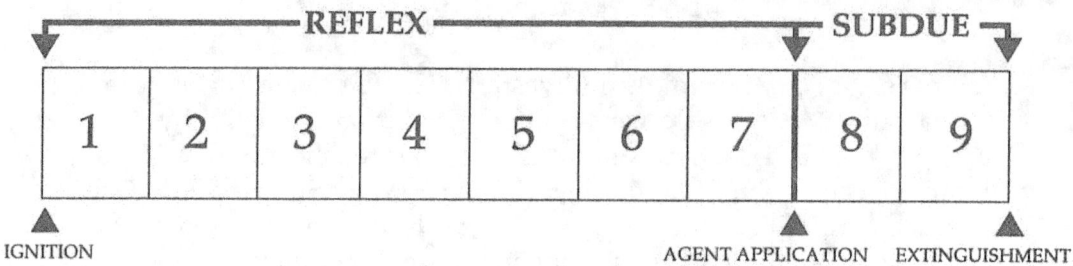

ACTION ON TIME ANALYSIS

Once the times are being gathered either by permanent collection or special study, they can be examined by Fire Service management to see what it can do to shorten the length of time to the **Arrival Point** once the alarm is received in the Alarm Center.

Likewise, the training officer and the incident commanders working together can develop means that would shorten the length of time to the **Agent Application Point** after the apparatus has arrived at the building fire incident scene.

EXAMPLE:

> The chief in one city decided that 58 calls for outside assistance each year couldn't be what his Fire Service actually required: In two cases, he noticed that the fires were being lost after his engines had arrived and **Agent Application** had begun.

> He worked with the training officer to develop an improved program for initial attack procedures using preconnected lines. He taught his officers to utilize aggressive "quash" techniques followed by vigorous initial interior attack if needed.

> The number of calls for outside assistance began to drop. Within four years, the total was down to only 4 calls for outside assistance from this Fire Service. In each of those four cases, the fire was at a size that required outside assistance before his first staff arrived.

> The time analysis this chief used was *common sense*.

There is no "right" way to use the nine defined **Steps** and the combined **Reflex Time** and **Subdue Time**. Some personnel will use instinct to show ways they can improve. Other Fire Services will provide a formal process for time study and management, and there will be many in between.

Analysis of the time periods given here has been found useful in thinking about the problem and may assist Fire Service officers in improving the quality and performance of their service.

Chapter 7
Summary

This chapter will give the reader:

- A review of the 9 Steps.

- A review of the 2 Key Times.

- A summary of the need for effective Agent Application.

"The object of the world of ideas is not the portrayal of reality — this would be utterly impossible — but rather to provide us with an instrument for finding our way about in this world more easily." — Professor Vaihinger, 1876

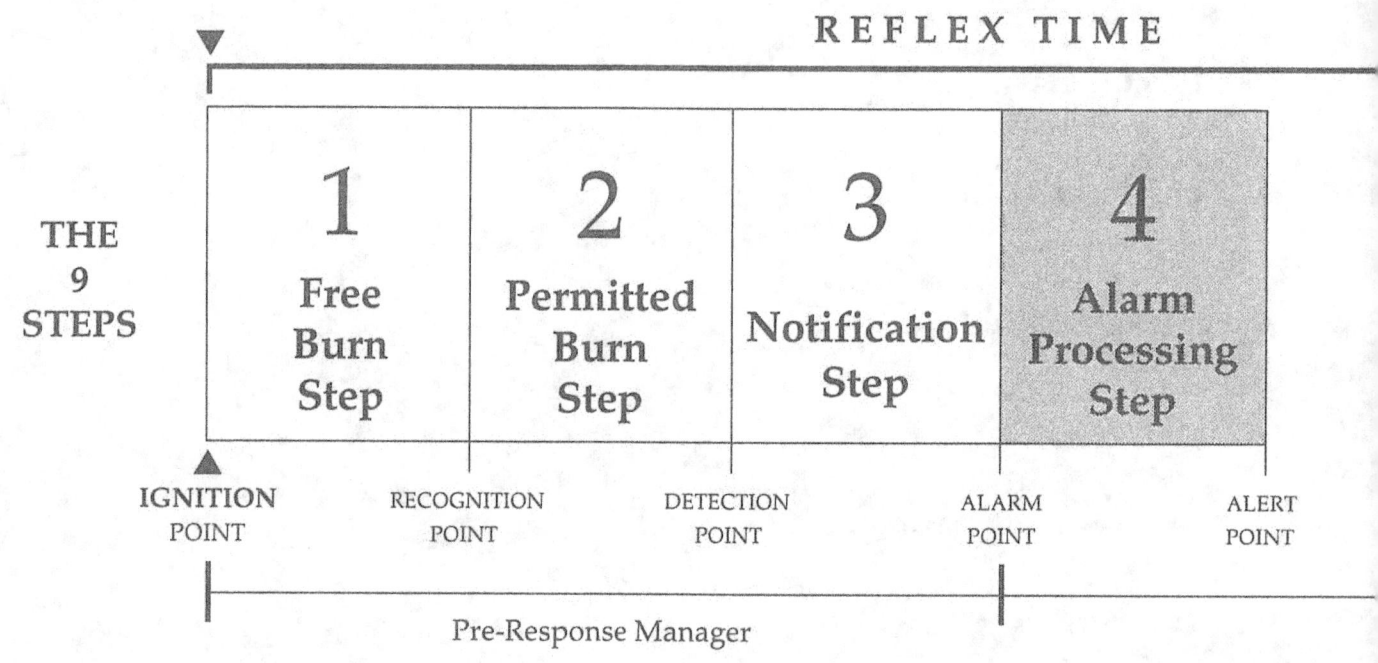

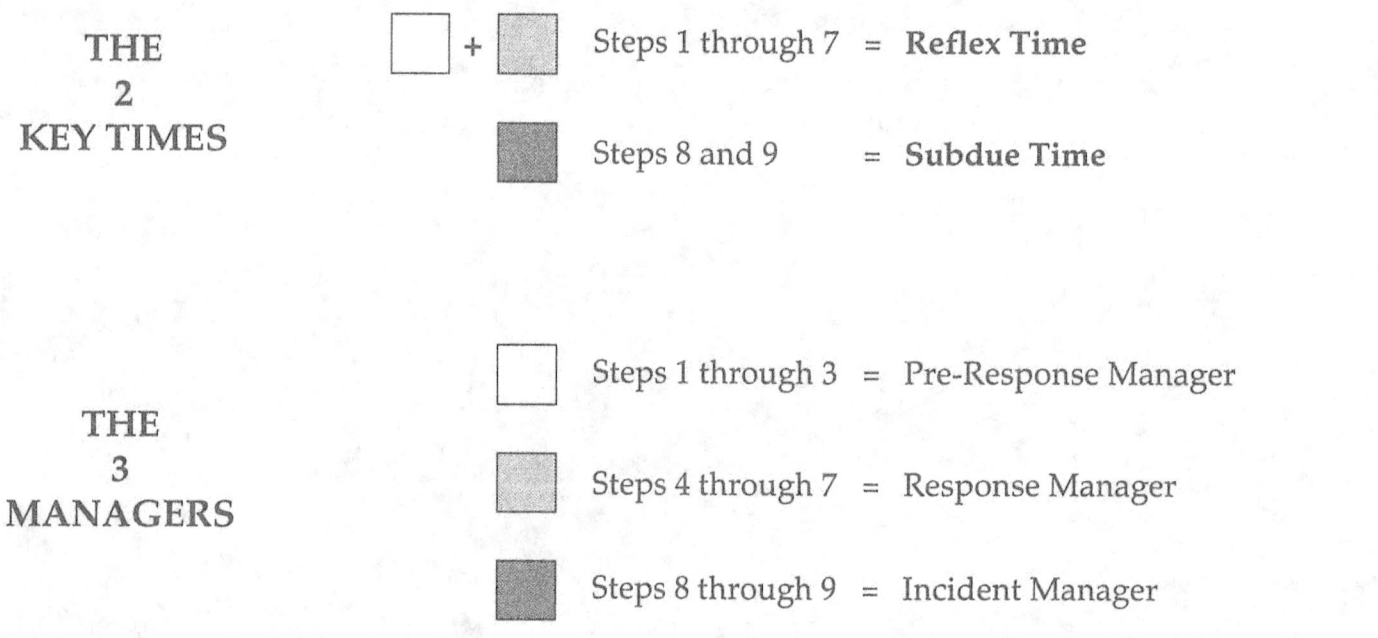

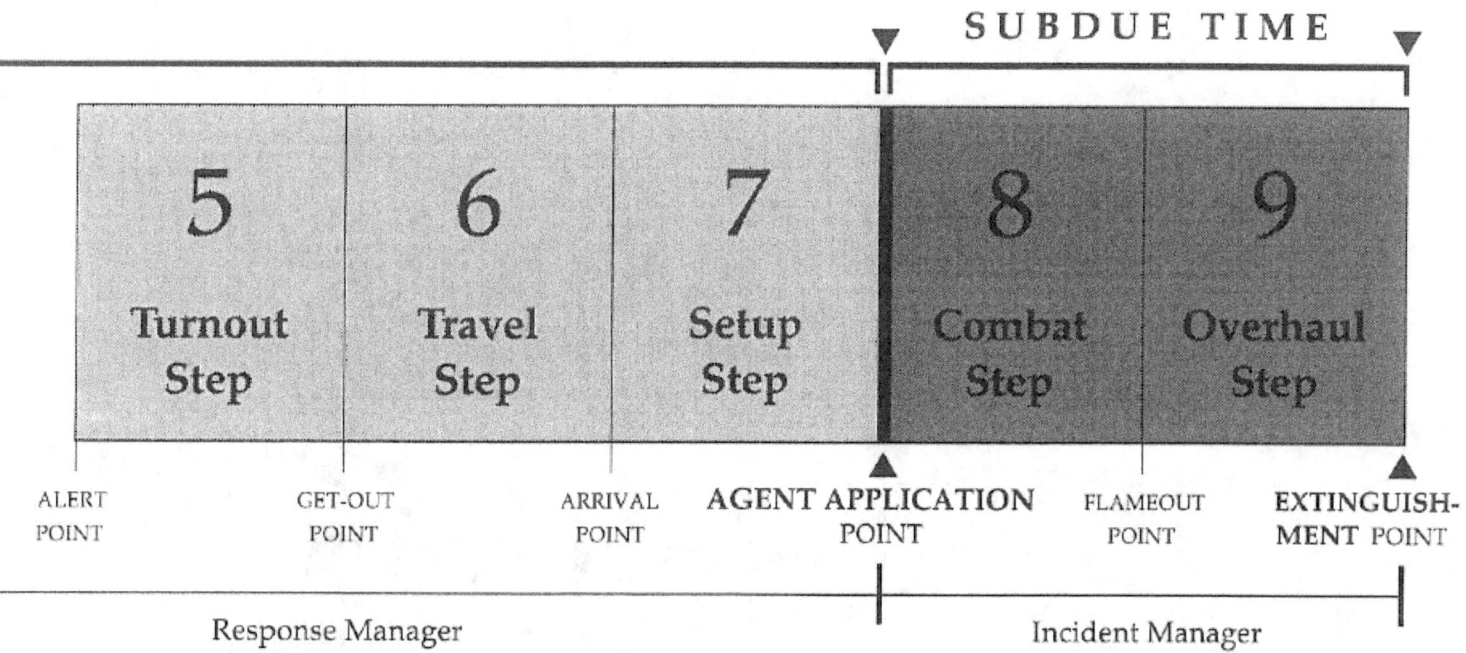

EXAMPLE: *[repeated for emphasis]*

In the city with a one-second **Alarm Processing Step**, a call came in from a frantic woman living at 50 Elm Street. In an excited voice with a thick accent, she reported that some kids had set fire to the outside of her house. The entire department heard her call.

The first-due engine company, the first-due ladder, and the first-due battalion chief — released on this initial verbal alarm — arrived at least 60 seconds before traditional **Alarm Processing** methods would have permitted. (Their home station was just 2-1/2 blocks from the caller's address.) The fire had burned through the outside sheathing and was beginning to travel up the open stud space.

The ladder company opened up the outside wall. Using a preconnected line, the engine company extinguished the fire 1.4 metres (4-1/2 feet) up the wall stud space.

This particular house was built using balloon construction methods. The time from the flame being 1.4 metres (4-1/2 feet) up the stud space to the flame entering the attic would have been less than 15 seconds by most estimates.

Clearly, the saving of 60 seconds in this case prevented major damage to the woman's house. It made the incident a "small fire" instead of the large, two-alarm attic fire most traditional Fire Service jurisdictions would have had to report.

SUMMARY

Fire grows at an exponential rate. Flames get worse every second.

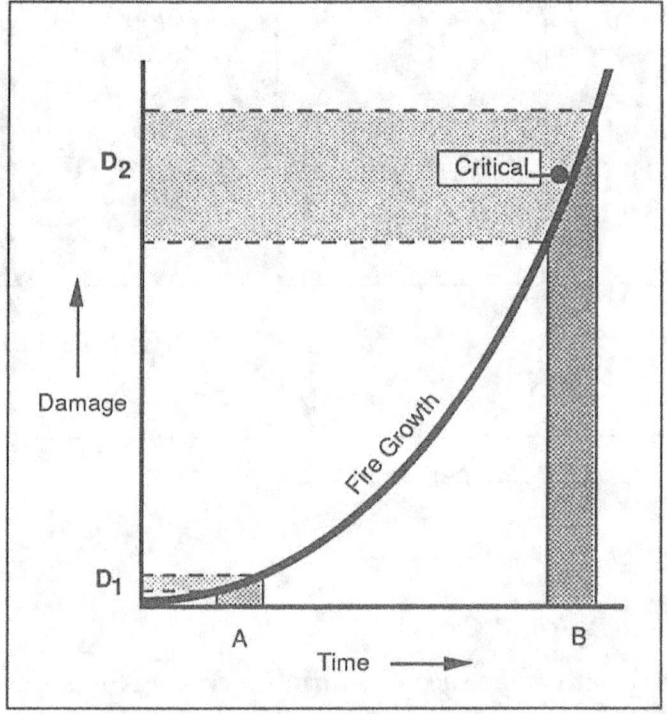

Figure 7-1. *Fire Growth is Exponential.*

Different ways of looking at operations — some calling for changes in traditional operational methods — can reduce time usage in manual fire suppression by a Fire Service.

The seconds saved in each step of **Reflex Time** are added together to reduce the total time elapsed before the "magic moment" of **Agent Application**.

The seconds saved before **Agent Application** result in a smaller fire being attacked.

Smaller fires on attack mean higher success rates, less severe losses, and, in some cases, saving human life.

INDEX

About the Institute

FIREPRO Institute Ltd., an international membership organization, was founded on the belief that all of us — worldwide — working together can find a simplified method to allow:

- the Building Community
  to specify what level of firesafety it wants, if any, and

- the Fire Community
  to deliver that level of firesafety.

The Institute is working to create a more effective basis for understanding between fire and building interests. Annual membership includes a subscription to *The Firesafety Designer* quarterly newsletter, as well as advance notice and discounts on publications and courses. For further information, contact FIREPRO Institute Ltd. through one of the following means:

Mail:	Postal Box 600; Putney, Vermont 05346; USA
Telephone:	(802) 387-5600
Facsimile:	(802) 387-4419
Internet E-Mail:	FIREPRO@aol.com

About the Author

Rexford Wilson earned his BSEE from the University of Massachusetts while serving on the Amherst (Massachusetts, USA) Fire Department. He joined the Factory Insurance Association (now IRI) as an inspector of American industry before being invited to the National Fire Protection Association. At the NFPA, he held posts as *Fire Record* editor, general field representative, and executive secretary of the Fire Marshals Association of North America.

Wilson became head of Fire Service Extension, a department of the College of Engineering at the University of Maryland, where he was responsible for training 15,000 in-state fire and rescue personnel. In 1970, he founded FIREPRO Incorporated and began a firesafety consulting career.

For contributions as a professional engineer, he has been awarded a patent for the application of Halon 1301, the NFPA Standards Medal, and the Society of Fire Protection Engineers' (SFPE) "Hats Off" Award.

A noted author and lecturer, Wilson co-founded the course on Measurement of Building Firesafety, founded the home fire management program, the NFPA Fire Reporting Committee, the SFPE Measurement of Fire Phenomena Committee, and the Zero Fires Program for industry.

He was elected a Fellow of the SFPE and served as its international secretary-treasurer. He has also been president of the New England Chapter of SFPE and was editor of its newsletter, *The Firebrand*.